T5-CQC-557

TENSOR ANALYSIS

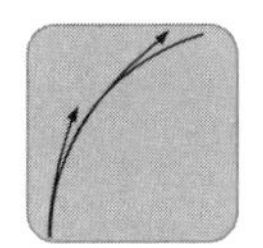

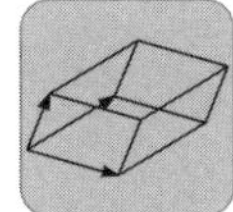

TENSOR ANALYSIS

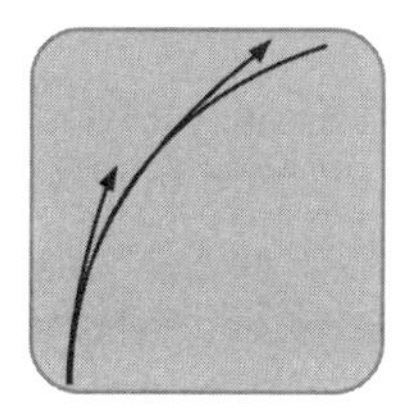

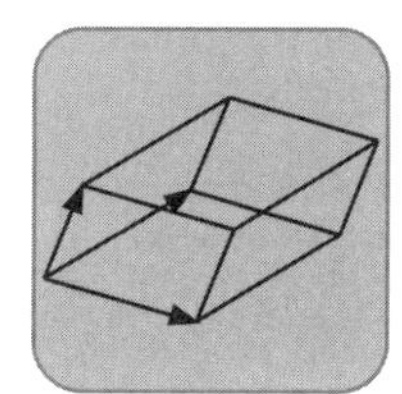

Leonid P Lebedev

National University of Colombia, Colombia
and Rostov State University, Russia

Michael J Cloud

Lawrence Technological University, USA

World Scientific

New Jersey • London • Singapore • Hong Kong

Published by

World Scientific Publishing Co. Pte. Ltd.
5 Toh Tuck Link, Singapore 596224
USA office: 27 Warren Street, Suite 401-402, Hackensack, NJ 07601
UK office: 57 Shelton Street, Covent Garden, London WC2H 9HE

British Library Cataloguing-in-Publication Data
A catalogue record for this book is available from the British Library.

First published 2003
Reprinted 2006

TENSOR ANALYSIS

ISBN 981-238-360-3

Printed in Singapore by World Scientific Printers (S) Pte Ltd

Foreword

Every science elaborates tools for the description of its objects of study. In classical mechanics we make extensive use of vectorial quantities: forces, moments, positions, velocities, momenta. Confining ourselves to a single coordinate frame, we can regard a vector as a fixed column matrix. The definitive trait of a vector quantity, however, is its objectivity; a vector does not depend on our choice of coordinate frame. This means that as soon as the components of a force are specified in one frame, the components of that force relative to any other frame can be found through the use of appropriate transformation rules.

But vector quantities alone do not suffice for the description of continuum media. The stress and strain at a point inside a body are also objective quantities; however, the specification of each of these relative to a given frame requires a square matrix of elements. Under changes of frame these elements transform according to rules different from those mentioned above. Stress and strain are examples of tensors of the second rank. We could go on to cite other objective quantities that occur in the mechanics of continua. The set of elastic moduli associated with Hooke's law comprise a tensor of the fourth rank; as such, these moduli obey yet another set of transformation rules. Despite the differences that exist between the transformation laws for the various types of objective quantities, they all fit into a unified scheme: the theory of tensors.

Tensor theory not only relieves our memory from a huge burden, but enables us to carry out differential operations with ease. This is the case even in curvilinear coordinate systems. Through the unmatched simplicity and brevity it affords, tensor analysis has attained the status of a general language that can be spoken across the various areas of continuum physics. A full comprehension of this language has become necessary for those working in electromagnetism, the theory of relativity, or virtually any other field-theoretic discipline. More modern books on physical subjects invariably contain appendices in which various vector and tensor identities are listed. These may suffice when one wishes to verify the steps in a development, but can leave one in doubt as to how the facts were established or, *a fortiori*, how they could be adapted to other circumstances. On the other hand, a comprehensive treatment of tensors (e.g., involving a full excursion into multilinear algebra) is necessarily so large as to be flatly inappropriate for the student or practicing engineer.

Hence the need for a treatment of tensor theory that does justice to the subject and is friendly to the practitioner. The authors of the present book have met these objectives with a presentation that is simple, clear, and sufficiently detailed. The concise text explains practically all those formulas needed to describe objects in three-dimensional space. Occurrences in physics are mentioned when helpful, but the discussion is kept largely independent of application area in order to appeal to the widest possible audience. A final chapter on the properties of curves and surfaces has been included; a brief introduction to the study of these properties can be considered as an informative and natural extension of tensor theory.

I.I. Vorovich
Professor of Mechanics and Mathematics
Rostov State University, Russia
Fellow of Russian Academy of Sciences

Preface

Originally a vector was regarded as an arrow of a certain length that could represent a force acting on a material point. Over a period of many years, this naive viewpoint evolved into the modern interpretation of the notion of vector and its extension to tensors. It was found that the use of vectors and tensors led to a proper description of certain properties and behaviors of real natural objects: those aspects that do not depend on the coordinate systems we introduce in space. This independence means that if we define such properties using one coordinate system, then in another system we can recalculate these characteristics using valid transformation rules. The ease with which a given problem can be solved often depends on the coordinate system employed. So in applications we must apply various coordinate systems, derive corresponding equations, and understand how to recalculate results in other systems. This book provides the tools necessary for such calculation.

Many physical laws are cumbersome when written in coordinate form but become compact and attractive looking when written in tensorial form. Such compact forms are easy to remember, and can be used to state complex physical boundary value problems. It is conceivable that soon an ability to merely formulate statements of boundary value problems will be regarded as a fundamental skill for the practitioner. Indeed, computer software is slowly advancing toward the point where the only necessary input data will be a coordinate-free statement of a boundary value problem; presumably the user will be able to initiate a solution process in a certain frame and by a certain method (analytical, numerical, or mixed), or simply ask the computer algorithm to choose the best frame and method. In this way vectors and tensors will become important elements of the macro-language for the next generation of software in engineering and applied mathematics.

We would like to thank the editorial staff at World Scientific — especially Mr. Kwang-Wei Tjan and Ms. Sook-Cheng Lim — for their assistance in the production of this book. Professor Byron C. Drachman of Michigan State University commented on the manuscript in its initial stages. Lastly, Natasha Lebedeva and Beth Lannon-Cloud deserve thanks for their patience and support.

Department of Mechanics and Mathematics L.P. Lebedev
Rostov State University, Russia

&

Department of Mathematics
National University of Colombia, Colombia

Department of Electrical and Computer Engineering M.J. Cloud
Lawrence Technological University, USA

Contents

Chapter 1

Preliminaries

1.1 The Vector Concept Revisited

The concept of a vector has been one of the most fruitful ideas in all of mathematics, and it is not surprising that we receive repeated exposure to the idea throughout our education. Students in elementary mathematics deal with vectors in component form — with quantities such as

$$\mathbf{x} = (2, 1, 3)$$

for example. But let us examine this situation more closely. Do the components $2, 1, 3$ determine the vector $\mathbf{x}$? They surely do if we specify the basis vectors of the coordinate frame. In elementary mathematics these are supposed to be mutually orthogonal and of unit length; even then they are not fully characterized, however, because such a frame can be rotated. In the description of many common phenomena we deal with vectorial quantities like forces that have definite directions and magnitudes. An example is the force your body exerts on a chair as you sit in front of the television set. This force does not depend on the coordinate frame employed by someone writing a textbook on vectors somewhere in Russia or China. Because the vector $\mathbf{f}$ representing a particular force is something objective, we should be able to write it in such a form that it ceases to depend on the details of the coordinate frame. The simplest way is to incorporate the frame vectors $\mathbf{e}_i$, $i = 1, 2, 3$, explicitly into the notation: if $\mathbf{x}$ is a vector we may write

$$\mathbf{x} = \sum_{i=1}^{3} x_i \mathbf{e}_i. \tag{1.1}$$

Then if we wish to change the frame, we should do so in such a way that $\mathbf{x}$ remains the same. This of course means that we cannot change only the frame vectors $\mathbf{e}_i$: we must change the components x_i correspondingly. So the components of a vector $\mathbf{x}$ in a new frame are not independent of those in the old frame.

1.2 A First Look at Tensors

In what follows we shall discuss how to work with vectors using different coordinate frames. Let us note that in mechanics there are objects of another nature. For example, there is a so-called tensor of inertia. This is an objective characteristic of a solid body, determining how the body rotates when torques act upon it. If the body is considered in a Cartesian frame, the tensor of inertia is described by a 3×3 matrix. If we change the frame, the matrix elements change according to certain rules. In textbooks on mechanics the reader can find lengthy discussions on how to change the matrix elements to maintain the same objective characteristic of the body when the new frame is also Cartesian. Although the tensor of inertia is objective (i.e., frame-independent), it is not a vector: it belongs to another class of mathematical objects. Many such *tensors of the second rank* arise in continuum mechanics: tensors of stress, strain, etc. They characterize certain properties of a body at each point; again, their "components" should transform in such a way that the tensors themselves do not depend on the frame.

For both vectors and tensors we can introduce various operations. Of course, the introduction of any new operation should be done in such a way that the results agree with known special cases when such familiar cases are met. If we introduce, say, dot multiplication of a tensor by a vector, then in a Cartesian frame the operation should resemble the multiplication of a matrix by a column vector. Similarly, the multiplication of two tensors should be defined so that in a Cartesian frame the operation involves matrix multiplication. To this end we consider *dyads* of vectors. These are quantities of the form[1]

$$\mathbf{e}_i \mathbf{e}_j .$$

[1]The quantity $\mathbf{e}_i\mathbf{e}_j$ is also called the *tensor product* of the vectors $\mathbf{e}_i$ and $\mathbf{e}_j$, and is sometimes denoted $\mathbf{e}_i \otimes \mathbf{e}_j$. Our notation (without the symbol $\otimes$) emphasizes that, for example, $\mathbf{e}_1\mathbf{e}_2$ is an elemental object belonging to the set of second-rank tensors, in the same way that $\mathbf{e}_1$ is an elemental object belonging to the set of vectors. Note that $\mathbf{e}_2\mathbf{e}_1$ and $\mathbf{e}_1\mathbf{e}_2$ are different objects, however. The term "tensor product" indicates that the operation shares certain properties with the product we know from elementary algebra.

A tensor may then be represented as

$$\sum_{i,j} a_{ij}\,\mathbf{e}_i\mathbf{e}_j$$

where the a_{ij} are the components of the tensor. We compare with equation (1.1) and notice the similarity in notation.

Natural objects can possess characteristics described by tensors of higher rank. For example, the elastic properties of a body are described by a tensor of the fourth rank (i.e., a tensor whose elemental parts are of the form $\mathbf{abcd}$, where $\mathbf{a}, \mathbf{b}, \mathbf{c}, \mathbf{d}$ are vectors). This means that in general the properties of a body are given by a "table" consisting of $3 \times 3 \times 3 \times 3 = 81$ elements. The elements change according to certain rules if we change the frame.

Tensors also occur in electrodynamics, the general theory of relativity, and many other sciences that deal with objects situated or distributed in space.

1.3 Assumed Background

In what follows we suppose a familiarity with the dot and cross products and their expression in Cartesian frames. Recall that if $\mathbf{a}$ and $\mathbf{b}$ are vectors, then by definition

$$\mathbf{a} \cdot \mathbf{b} = |\mathbf{a}||\mathbf{b}| \cos\theta,$$

where $|\mathbf{a}|$ and $|\mathbf{b}|$ are the magnitudes of $\mathbf{a}$ and $\mathbf{b}$ and θ is the (smaller) angle between $\mathbf{a}$ and $\mathbf{b}$. In a Cartesian frame with basis vectors[2] $\mathbf{i}_1, \mathbf{i}_2, \mathbf{i}_3$ where $\mathbf{a}$ and $\mathbf{b}$ are expressed as

$$\mathbf{a} = a_1\mathbf{i}_1 + a_2\mathbf{i}_2 + a_3\mathbf{i}_3, \qquad \mathbf{b} = b_1\mathbf{i}_1 + b_2\mathbf{i}_2 + b_3\mathbf{i}_3,$$

we have

$$\mathbf{a} \cdot \mathbf{b} = a_1b_1 + a_2b_2 + a_3b_3.$$

Also recall that

$$\mathbf{a} \times \mathbf{b} = \begin{vmatrix} \mathbf{i}_1 & \mathbf{i}_2 & \mathbf{i}_3 \\ a_1 & a_2 & a_3 \\ b_1 & b_2 & b_3 \end{vmatrix}.$$

[2]From now on we reserve the symbol $\mathbf{i}$ for the basis vectors of a Cartesian system.

The dot product will play a role in our discussion from the very beginning. The cross product will be used as needed, and a fuller discussion will appear in § 2.6.

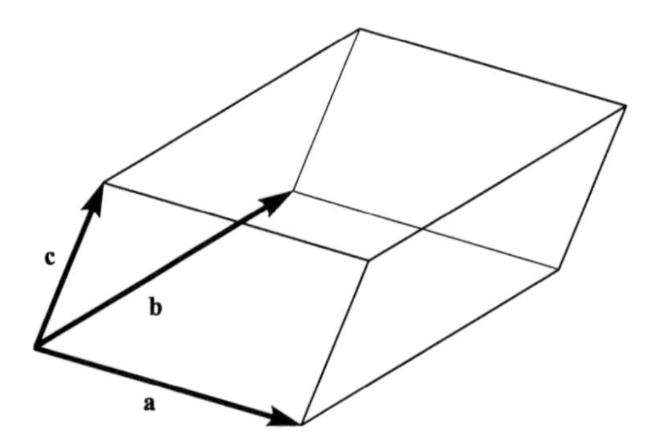

Fig. 1.1 Geometrical meaning of the scalar triple product.

Given three vectors $\mathbf{a}, \mathbf{b}, \mathbf{c}$ we can form the *scalar triple product*

$$\mathbf{a} \cdot (\mathbf{b} \times \mathbf{c}).$$

This may be interpreted as the volume of the parallelepiped having $\mathbf{a}, \mathbf{b}, \mathbf{c}$ as three of its co-terminal edges (Fig. 1.1). In rectangular components we have, according to the expressions above,

$$\mathbf{a} \cdot (\mathbf{b} \times \mathbf{c}) = \begin{vmatrix} a_1 & a_2 & a_3 \\ b_1 & b_2 & b_3 \\ c_1 & c_2 & c_3 \end{vmatrix}.$$

Permissible manipulations with the scalar triple product include cyclic interchange of the vectors:

$$\mathbf{a} \cdot (\mathbf{b} \times \mathbf{c}) = \mathbf{c} \cdot (\mathbf{a} \times \mathbf{b}) = \mathbf{b} \cdot (\mathbf{c} \times \mathbf{a}).$$

Exercise 1.1 What does the condition $\mathbf{a} \cdot (\mathbf{b} \times \mathbf{c}) \neq 0$ say about $\mathbf{a}$, $\mathbf{b}$, and $\mathbf{c}$? (Hints for this and many other exercises appear in Appendix B beginning on page 165.)

So far we have made reference to vectors in a three-dimensional space, and shall continue this practice throughout most of the book. It is also possible (and sometimes useful) to introduce vectors in a more general space of $n > 3$ dimensions, e.g.,

$$\mathbf{a} = \mathbf{i}_1 a_1 + \mathbf{i}_2 a_2 + \cdots + \mathbf{i}_n a_n.$$

It turns out that many (but not all) of the principles and techniques we shall learn have direct extensions to such higher-dimensional spaces. It is also true that many three-dimensional notions can be reconsidered in *two* dimensions. The reader should take the time to reduce each three-dimensional formula to its two-dimensional analogue to understand what happens with the corresponding assertion.

1.4 More on the Notion of a Vector

Before closing out this chapter we should mention the notions of a vector as a "directed line segment" or "quantity having magnitude and direction." We find these in elementary math and physics textbooks. But it is easy to point to a situation in which a quantity of interest has magnitude and direction but is not a vector. The total electric current flowing in a thin wire is one example: to describe the current we must specify the rate of flow of electrons, the orientation of the wire, and the sense of electron movement along the wire. However, if two wires lie in a plane and carry equal currents running perpendicular to each other, then we cannot duplicate their physical effects by replacing them with a third wire directed at a $45°$ angle with respect to the first two. Total electric currents cannot be considered as vector quantities since they do not combine according to the rule for vector addition.

Another problem concerns the notion of an n-dimensional Euclidean space. We sometimes hear it defined as the set each point of which is uniquely determined by n parameters. However, it is not reasonable to regard every list of n things as a possible vector. A tailor may take measurements for her client and compose an ordered list of lengths, widths, etc., and by the above "definition" the set of all such lists is an n-dimensional space. But in $\mathbb{R}^n$ any two points are joined by a vector whose components equal the differences of the corresponding coordinates of the points. In the tailor's case this notion is completely senseless. To make matters worse, in $\mathbb{R}^n$ one can multiply any vector by any real number and obtain another vector in the space. A tailor's list showing that someone is six meters tall would be pretty hard to find.

Such simplistic definitions can be dangerous. The notion of a vector was eventually elaborated in physics, more precisely in mechanics where the behavior of forces was used as the model for a general vector. However, forces have some rather strange features: for example, if we shift the point of application of a force acting on a solid body, then we must introduce a moment acting on the body or the motion of the body will change. So the mechanical forces that gave birth to the notion of a vector possess features not covered by the mathematical definition of a vector. In the classical mechanics of a rigid body we are allowed to move a force along its line of action but cannot simply shift it off that line. With a deformable body we cannot move a force anywhere because in doing so we immediately change the state of the body. We can vectorially add two forces acting on a rigid body (not forgetting about the moments arising during shift of the forces). On the other hand, if two forces act on two different material points then we can add the vectors that represent the forces, but will not necessarily obtain a new vector that is relevant to the physics of the situation. So we should understand that the idea of a vector in mathematics reflects only some of the features of the real object it describes.

We would like to mention something else about vectors. In elementary mathematics students use vectors and points quite interchangeably. However, these are objects of different natures: there are vectors in space and there are the points of a space. We can, for instance, associate with a vector in an n-dimensional Euclidean vector space a point in an n-dimensional Euclidean point space. We can then consider a vector $\mathbf{x}$ as a shift of all points of the point space by an amount specified by $\mathbf{x}$. The result of this map is the same space of points; each pair of points that correspond under the mapping define a vector $\mathbf{x}$ under which a point shifts into its image. This vector is what we find when we subtract the Cartesian coordinates of the initial point from those of the final point. If we add the fact that the composition of two such maps obeys the rules of vector addition, then we get a strict definition of the space introduced intuitively in elementary mathematics. Engineers might object to the imposition of such formality on an apparently simple situation, but mathematical rigor has proved its worth from many practical points of view. For example, a computer that processes information in the complete absence of intuition can deal properly with objects for which rigorous definitions and manipulation rules have been formulated.

This brings us to our final word, regarding the expression of vectors in component-free notation. The simple and compact notation $\mathbf{x}$ for a vector leads to powerful ways of exhibiting relationships between the many vector (and tensor) quantities that occur in mathematical physics. It permits us to accomplish manipulations that would take many pages and become quite confusing if done in component form. This is typical for problems of nonlinear physics, and for those where change of coordinate frame becomes necessary. The resulting formal nature of the manipulations means that computers can be expected to take on more and more of this sort of work, even at the level of original research.

Chapter 2

Transformations and Vectors

2.1 Change of Basis

Let us reconsider the vector $\mathbf{x} = (2, 1, 3)$. Fully written out, it is

$$\mathbf{x} = 2\mathbf{e}_1 + \mathbf{e}_2 + 3\mathbf{e}_3$$

in a given Cartesian[1] frame $\mathbf{e}_i$, $i = 1, 2, 3$. Suppose we appoint a new frame $\tilde{\mathbf{e}}_i$, $i = 1, 2, 3$, such that

$$\begin{aligned}
\mathbf{e}_1 &= \tilde{\mathbf{e}}_1 + 2\tilde{\mathbf{e}}_2 + 3\tilde{\mathbf{e}}_3, \\
\mathbf{e}_2 &= 4\tilde{\mathbf{e}}_1 + 5\tilde{\mathbf{e}}_2 + 6\tilde{\mathbf{e}}_3, \\
\mathbf{e}_3 &= 7\tilde{\mathbf{e}}_1 + 8\tilde{\mathbf{e}}_2 + 9\tilde{\mathbf{e}}_3.
\end{aligned}$$

(From these expansions we could calculate the $\tilde{\mathbf{e}}_i$ and verify that they are non-coplanar.) Recalling that $\mathbf{x}$ is an objective, frame-independent entity, we can write

$$\begin{aligned}
\mathbf{x} &= 2(\tilde{\mathbf{e}}_1 + 2\tilde{\mathbf{e}}_2 + 3\tilde{\mathbf{e}}_3) + (4\tilde{\mathbf{e}}_1 + 5\tilde{\mathbf{e}}_2 + 6\tilde{\mathbf{e}}_3) + 3(7\tilde{\mathbf{e}}_1 + 8\tilde{\mathbf{e}}_2 + 9\tilde{\mathbf{e}}_3) \\
&= (2 + 4 + 21)\tilde{\mathbf{e}}_1 + (4 + 5 + 24)\tilde{\mathbf{e}}_2 + (6 + 6 + 27)\tilde{\mathbf{e}}_3 \\
&= 27\tilde{\mathbf{e}}_1 + 33\tilde{\mathbf{e}}_2 + 39\tilde{\mathbf{e}}_3.
\end{aligned}$$

In these calculations it is unimportant whether the frames are Cartesian; it is important only that we have the table of transformation

$$\begin{pmatrix} 1 & 2 & 3 \\ 4 & 5 & 6 \\ 7 & 8 & 9 \end{pmatrix}.$$

[1] This is one of the few times in which we do not use $\mathbf{i}$ as the symbol for a Cartesian frame vector.

It is clear that we can repeat the same operation in general form. Let $\mathbf{x}$ be of the form

$$\mathbf{x} = \sum_{i=1}^{3} x^i \mathbf{e}_i \tag{2.1}$$

with the table of transformation of the frame given as

$$\mathbf{e}_i = \sum_{j=1}^{3} A_i^j \tilde{\mathbf{e}}_j.$$

Then

$$\mathbf{x} = \sum_{i=1}^{3} x^i \sum_{j=1}^{3} A_i^j \tilde{\mathbf{e}}_j = \sum_{j=1}^{3} \tilde{\mathbf{e}}_j \sum_{i=1}^{3} A_i^j x^i.$$

Here have introduced a new notation, placing some indices as subscripts and some as superscripts. Although this practice may seem artificial, there are some fairly deep reasons for following it.

2.2 Dual Bases

To perform operations with a vector $\mathbf{x}$ we must have a straightforward method of calculating its components — ultimately, no matter how advanced we are, we must be able to obtain the x^i using simple arithmetic. We prefer formulas that permit us to find the components of vectors using dot multiplication only; we shall need these when doing frame transformations, etc. In a Cartesian frame the necessary operation is simple dot multiplication by the corresponding basis vector of the frame: we have

$$x^k = \mathbf{x} \cdot \mathbf{i}_k, \qquad k = 1, 2, 3.$$

It is clear that this same procedure will fail in a more general non-Cartesian frame (since we will not necessarily have $\mathbf{e}_i \cdot \mathbf{e}_j = 0$ for all $j \neq i$). However, it may still be possible to find some vector, call it $\mathbf{e}^i$, such that

$$x^i = \mathbf{x} \cdot \mathbf{e}^i, \qquad i = 1, 2, 3$$

in this more general situation. If we set

$$x^i = \mathbf{x} \cdot \mathbf{e}^i = \left(\sum_{j=1}^{3} x^j \mathbf{e}_j \right) \cdot \mathbf{e}^i = \sum_{j=1}^{3} x^j (\mathbf{e}_j \cdot \mathbf{e}^i)$$

and compare the left- and right-hand sides, we see that equality holds when

$$\mathbf{e}_j \cdot \mathbf{e}^i = \delta_j^i \tag{2.2}$$

where

$$\delta_j^i = \begin{cases} 1, & j = i, \\ 0, & j \neq i, \end{cases}$$

is the *Kronecker delta* symbol. In a Cartesian frame we have $\mathbf{e}^k = \mathbf{e}_k = \mathbf{i}_k$ for each k.

Exercise 2.1 Show that $\mathbf{e}^i$ is determined uniquely by the requirement that $x^i = \mathbf{x} \cdot \mathbf{e}^i$ for every $\mathbf{x}$.

Now let us discuss the geometrical nature of the vectors $\mathbf{e}^i$. Consider, for example, the equations for $\mathbf{e}^1$:

$$\mathbf{e}_1 \cdot \mathbf{e}^1 = 1, \qquad \mathbf{e}_2 \cdot \mathbf{e}^1 = 0, \qquad \mathbf{e}_3 \cdot \mathbf{e}^1 = 0.$$

We see that $\mathbf{e}^1$ is orthogonal to both $\mathbf{e}_2$ and $\mathbf{e}_3$, and its magnitude is such that $\mathbf{e}_1 \cdot \mathbf{e}^1 = 1$. Similar properties hold for $\mathbf{e}^2$ and $\mathbf{e}^3$.

Exercise 2.2 Show that the vectors $\mathbf{e}^i$ are linearly independent.

By the last exercise the $\mathbf{e}^i$ constitute a frame or basis. We say that this basis is *reciprocal* or *dual* to the basis $\mathbf{e}_i$. We can therefore expand an arbitrary vector $\mathbf{x}$ as

$$\mathbf{x} = \sum_{i=1}^{3} x_i \mathbf{e}^i. \tag{2.3}$$

Note that superscripts and subscripts continue to appear in our notation, but in a way complementary to that used in equation (2.1). If we dot-multiply the representation (2.3) of $\mathbf{x}$ by $\mathbf{e}_j$ and use (2.2) we get x_j. This explains why the frames $\mathbf{e}^i$ and $\mathbf{e}_i$ are dual: the formulas

$$\mathbf{x} \cdot \mathbf{e}^i = x^i, \qquad \mathbf{x} \cdot \mathbf{e}_i = x_i,$$

look quite similar. So the introduction of a reciprocal basis gives many potential advantages.

Let us discuss the reciprocal basis in more detail. The first problem is to find suitable formulas to define it. We derive these formulas next, but first let us note the following. The use of reciprocal vectors may not be practical in those situations where we are working with only two or three vectors. The real advantages come when we are working intensively with many vectors. This is reminiscent of the solution of a set of linear simultaneous equations: it is inefficient to find the inverse matrix of the system if we have only one forcing vector. But when we must solve such a problem repeatedly for many forcing vectors, the calculation and use of the inverse matrix becomes sensible.

Writing out $\mathbf{x}$ in the $\mathbf{e}_i$ and $\mathbf{e}^i$ bases, we used a combination of indices (i.e., subscripts and superscripts) and summation symbols. From now on we shall omit the symbol of summation when we meet matching subscripts and superscripts: we shall write, say, $x_i a^i$ for the sum $\sum_i x_i a^i$. That is, whenever we see i as a subscript and a superscript, we shall understand that a summation is to be carried out over i. This rule shall apply to situations involving vectors as well: we shall understand, for example, $x^i \mathbf{e}_i$ to mean the summation $\sum_i x^i \mathbf{e}_i$. This rule is called the *rule of summation over repeated indices.*[2] Note that a repeated index is a *dummy index* in the sense that it may be replaced by any other index not already in use: we have

$$x_i a^i = x_1 a^1 + x_2 a^2 + x_3 a^3 = x_k a^k$$

for instance. An index that occurs just once in an expression, for example the index i in

$$A_i^k x_k,$$

is called a *free index.* In tensor discussions each free index is understood to range independently over a set of values — presently this set is $\{1, 2, 3\}$.

Let us return to the task of deriving formulas for the reciprocal basis vectors $\mathbf{e}^i$ in terms of the original basis vectors $\mathbf{e}_i$. We construct $\mathbf{e}^1$ first. Since the cross product of two vectors is perpendicular to both, we can satisfy the conditions $\mathbf{e}_2 \cdot \mathbf{e}^1 = 0$ and $\mathbf{e}_3 \cdot \mathbf{e}^1 = 0$ by setting

$$\mathbf{e}^1 = c_1(\mathbf{e}_2 \times \mathbf{e}_3)$$

where c_1 is a constant. To determine c_1 we impose the remaining condition $\mathbf{e}_1 \cdot \mathbf{e}^1 = 1$:

$$c_1[\mathbf{e}_1 \cdot (\mathbf{e}_2 \times \mathbf{e}_3)] = 1.$$

[2]The rule of summation was first introduced not by mathematicians but by Einstein, and is sometimes referred to as the "Einstein summation convention." In a paper where he introduced this rule, Einstein used Cartesian frames and therefore did not distinguish superscripts from subscripts. However, we shall continue to make the distinction so that we can deal with non-Cartesian frames.

The quantity $\mathbf{e}_1 \cdot (\mathbf{e}_2 \times \mathbf{e}_3)$ is a scalar that geometrically represents the volume of the parallelepiped described by the vectors $\mathbf{e}_i$. Denoting it by V, we have

$$\mathbf{e}^1 = \frac{1}{V}(\mathbf{e}_2 \times \mathbf{e}_3).$$

Similarly,

$$\mathbf{e}^2 = \frac{1}{V}(\mathbf{e}_3 \times \mathbf{e}_1), \qquad \mathbf{e}^3 = \frac{1}{V}(\mathbf{e}_1 \times \mathbf{e}_2).$$

The reader may verify that these expressions satisfy (2.2). Let us mention that if we construct the reciprocal basis to the basis $\mathbf{e}^i$ we obtain the initial basis $\mathbf{e}_i$. Hence we immediately get the dual formulas

$$\mathbf{e}_1 = \frac{1}{V'}(\mathbf{e}^2 \times \mathbf{e}^3), \qquad \mathbf{e}_2 = \frac{1}{V'}(\mathbf{e}^3 \times \mathbf{e}^1), \qquad \mathbf{e}_3 = \frac{1}{V'}(\mathbf{e}^1 \times \mathbf{e}^2),$$

where $V' = \mathbf{e}^1 \cdot (\mathbf{e}^2 \times \mathbf{e}^3)$ is the volume of the parallelepiped described by the vectors $\mathbf{e}^i$.

Exercise 2.3 Show that $V' = 1/V$.

Let us now consider the forms of the dot product of $\mathbf{a} = a^i\mathbf{e}_i = a_j\mathbf{e}^j$ with $\mathbf{b} = b^p\mathbf{e}_p = b_q\mathbf{e}^q$:

$$\mathbf{a} \cdot \mathbf{b} = a^i\mathbf{e}_i \cdot b^p\mathbf{e}_p = a^i b^p \mathbf{e}_i \cdot \mathbf{e}_p.$$

Introducing the notation

$$g_{ip} = \mathbf{e}_i \cdot \mathbf{e}_p, \tag{2.4}$$

we have

$$\mathbf{a} \cdot \mathbf{b} = a^i b^p g_{ip}.$$

(As a short exercise the reader should write out this expression in full.) Using the reciprocal component representations we get

$$\mathbf{a} \cdot \mathbf{b} = a_j\mathbf{e}^j \cdot b_q\mathbf{e}^q = a_j b_q g^{jq}$$

where

$$g^{jq} = \mathbf{e}^j \cdot \mathbf{e}^q. \tag{2.5}$$

Finally, using a mixed representation we get

$$\mathbf{a} \cdot \mathbf{b} = a^i\mathbf{e}_i \cdot b_q\mathbf{e}^q = a^i b_q \delta_i^q = a^i b_i$$

and, similarly,

$$\mathbf{a} \cdot \mathbf{b} = a_j b^j.$$

Hence

$$\mathbf{a} \cdot \mathbf{b} = a^i b^j g_{ij} = a_i b_j g^{ij} = a^i b_i = a_i b^i.$$

We see that when we use mixed bases to represent **a** and **b** we get formulas that resemble the equation

$$\mathbf{a} \cdot \mathbf{b} = a_1 b_1 + a_2 b_2 + a_3 b_3$$

from § 1.3; otherwise we get more terms and additional multipliers. Later we shall meet g_{ij} and g^{ij} very frequently. They are the components of a unique tensor known as the *metric tensor*. In Cartesian frames we obviously have $g_{ij} = \delta_i^j$ and $g^{ij} = \delta_j^i$.

2.3 Transformation to the Reciprocal Frame

How do the components of a vector **x** transform when we change to the reciprocal frame? We simply set

$$x_i \mathbf{e}^i = x^i \mathbf{e}_i$$

and dot both sides with $\mathbf{e}_j$ to get

$$x_i \mathbf{e}^i \cdot \mathbf{e}_j = x^i \mathbf{e}_i \cdot \mathbf{e}_j$$

or

$$x_j = x^i g_{ij}. \tag{2.6}$$

In the system of equations

$$\begin{pmatrix} x_1 \\ x_2 \\ x_3 \end{pmatrix} = \begin{pmatrix} g_{11} & g_{21} & g_{31} \\ g_{12} & g_{22} & g_{32} \\ g_{13} & g_{23} & g_{33} \end{pmatrix} \begin{pmatrix} x^1 \\ x^2 \\ x^3 \end{pmatrix}$$

the matrix of g_{ij} is called the *Gram matrix*. If its determinant is not zero then the vectors $\mathbf{e}_i$ are linearly independent.

Exercise 2.4 (a) Show that if the Gram determinant vanishes, then the $\mathbf{e}_i$ are linearly dependent. (b) Prove that the Gram determinant equals V^2.

We called the basis $\mathbf{e}_i$ dual to the basis $\mathbf{e}^i$. In $\mathbf{e}^i$ the metric components are given by g^{ij}, hence we can immediately write out a dual expression to (2.6):

$$x^i = x_j g^{ij}. \tag{2.7}$$

We see from (2.6) and (2.7) that, using the components of the metric tensor, we can always change subscripts to superscripts and vice versa. These actions are known as the *raising* and *lowering of indices.* Finally, (2.6) and (2.7) together imply

$$x_i = g_{ij} g^{jk} x_k,$$

hence

$$g_{ij} g^{jk} = \delta_i^k.$$

Of course, this means that the matrices of g_{ij} and g^{ij} are mutually inverse.

Quick summary

Given a basis $\mathbf{e}_i$, the vectors $\mathbf{e}^i$ given by the requirement that

$$\mathbf{e}_j \cdot \mathbf{e}^i = \delta_j^i$$

are linearly independent and form a basis called the reciprocal or dual basis. The definition of dual basis is motivated by the equation $x^i = \mathbf{x} \cdot \mathbf{e}^i$. Explicit expressions for the $\mathbf{e}^k$ can be written as

$$\mathbf{e}^i = \frac{1}{V}(\mathbf{e}_j \times \mathbf{e}_k)$$

where the ordered triple (i, j, k) equals $(1, 2, 3)$ or one of the cyclic permutations $(2, 3, 1)$ or $(3, 1, 2)$, and where $V = \mathbf{e}_1 \cdot (\mathbf{e}_2 \times \mathbf{e}_3)$ is the volume of the frame parallelepiped. The dual of the basis $\mathbf{e}^k$ (i.e., the dual of the dual) is the original basis $\mathbf{e}_k$. A given vector $\mathbf{x}$ can be expressed as

$$\mathbf{x} = x^i \mathbf{e}_i = x_i \mathbf{e}^i$$

where the x_i are the components of $\mathbf{x}$ with respect to the dual basis.

Exercise 2.5 (a) Let $\mathbf{x} = x^k \mathbf{e}_k = x_k \mathbf{e}^k$. Write out the modulus of $\mathbf{x}$ in all possible forms using the metric tensor. (b) Write out all forms of the dot product $\mathbf{x} \cdot \mathbf{y}$.

2.4 Transformation Between General Frames

Having transformed the components x^i of a vector $\mathbf{x}$ to the corresponding components x_i relative to the reciprocal basis, we are now ready to take on the more general task of transforming the x^i to the corresponding components $\tilde{x}^i$ relative to *any* other basis $\tilde{\mathbf{e}}_i$. Let the new basis $\tilde{\mathbf{e}}_i$ be related to the original basis $\mathbf{e}_i$ by

$$\mathbf{e}_i = A_i^j \tilde{\mathbf{e}}_j. \tag{2.8}$$

This is, of course, compact notation for the system of equations

$$\begin{pmatrix} \mathbf{e}_1 \\ \mathbf{e}_2 \\ \mathbf{e}_3 \end{pmatrix} = \underbrace{\begin{pmatrix} A_1^1 & A_1^2 & A_1^3 \\ A_2^1 & A_2^2 & A_2^3 \\ A_3^1 & A_3^2 & A_3^3 \end{pmatrix}}_{\equiv A, \text{ say}} \begin{pmatrix} \tilde{\mathbf{e}}_1 \\ \tilde{\mathbf{e}}_2 \\ \tilde{\mathbf{e}}_3 \end{pmatrix}.$$

Before proceeding, we note that in the symbol A_i^j the subscript indexes the row number in the matrix A, while the superscript indexes the column number. Throughout our development we shall often take the time to write various equations of interest in matrix notation. It follows from (2.8) that

$$A_i^j = \mathbf{e}_i \cdot \tilde{\mathbf{e}}^j.$$

Exercise 2.6 A Cartesian frame is rotated about its third axis to give a new Cartesian frame. Find the matrix of transformation.

A vector $\mathbf{x}$ can be expressed in the two forms

$$\mathbf{x} = x^k \mathbf{e}_k, \qquad \mathbf{x} = \tilde{x}^i \tilde{\mathbf{e}}_i.$$

Equating these two expressions for the same vector $\mathbf{x}$, we have

$$\tilde{x}^i \tilde{\mathbf{e}}_i = x^k \mathbf{e}_k,$$

hence

$$\tilde{x}^i \tilde{\mathbf{e}}_i = x^k A_k^j \tilde{\mathbf{e}}_j. \tag{2.9}$$

To find the $\tilde{x}^i$ in terms of the x^i, we may expand the notation and write (2.9) as

$$\tilde{x}^1 \tilde{\mathbf{e}}_1 + \tilde{x}^2 \tilde{\mathbf{e}}_2 + \tilde{x}^3 \tilde{\mathbf{e}}_3 = x^1 A_1^j \tilde{\mathbf{e}}_j + x^2 A_2^j \tilde{\mathbf{e}}_j + x^3 A_3^j \tilde{\mathbf{e}}_j$$

where, of course,

$$\begin{aligned} A_1^j \tilde{\mathbf{e}}_j &= A_1^1 \tilde{\mathbf{e}}_1 + A_1^2 \tilde{\mathbf{e}}_2 + A_1^3 \tilde{\mathbf{e}}_3, \\ A_2^j \tilde{\mathbf{e}}_j &= A_2^1 \tilde{\mathbf{e}}_1 + A_2^2 \tilde{\mathbf{e}}_2 + A_2^3 \tilde{\mathbf{e}}_3, \\ A_3^j \tilde{\mathbf{e}}_j &= A_3^1 \tilde{\mathbf{e}}_1 + A_3^2 \tilde{\mathbf{e}}_2 + A_3^3 \tilde{\mathbf{e}}_3. \end{aligned}$$

Matching coefficients of the $\tilde{\mathbf{e}}_i$ we find

$$\begin{aligned} \tilde{x}^1 &= x^1 A_1^1 + x^2 A_2^1 + x^3 A_3^1 = x^j A_j^1, \\ \tilde{x}^2 &= x^1 A_1^2 + x^2 A_2^2 + x^3 A_3^2 = x^j A_j^2, \\ \tilde{x}^3 &= x^1 A_1^3 + x^2 A_2^3 + x^3 A_3^3 = x^j A_j^3, \end{aligned}$$

hence

$$\tilde{x}^i = x^j A_j^i. \tag{2.10}$$

It is possible to obtain (2.10) from (2.9) in a succinct manner. On the right-hand side of (2.9) the index j is a dummy index which we can replace with i and thereby obtain (2.10) immediately. The matrix notation equivalent of (2.10) is

$$\begin{pmatrix} \tilde{x}^1 \\ \tilde{x}^2 \\ \tilde{x}^3 \end{pmatrix} = \begin{pmatrix} A_1^1 & A_2^1 & A_3^1 \\ A_1^2 & A_2^2 & A_3^2 \\ A_1^3 & A_2^3 & A_3^3 \end{pmatrix} \begin{pmatrix} x^1 \\ x^2 \\ x^3 \end{pmatrix}$$

and thus involves multiplication by A^T, the transpose of A.

We shall also need the equations of transformation from the frame $\tilde{\mathbf{e}}_i$ back to the frame $\mathbf{e}_i$. Since the direct transformation is linear the inverse must be linear as well, so we can write

$$\tilde{\mathbf{e}}_i = \tilde{A}_i^j \mathbf{e}_j \tag{2.11}$$

where

$$\tilde{A}_i^j = \tilde{\mathbf{e}}_i \cdot \mathbf{e}^j.$$

Let us find the relation between the matrices of transformation A and $\tilde{A}$. By (2.11) and (2.8) we have

$$\tilde{\mathbf{e}}_i = \tilde{A}_i^j \mathbf{e}_j = \tilde{A}_i^j A_j^k \tilde{\mathbf{e}}_k,$$

and since the $\tilde{\mathbf{e}}_i$ form a basis we must have

$$\tilde{A}_i^j A_j^k = \delta_i^k.$$

The relationship

$$A_i^j \tilde{A}_j^k = \delta_i^k$$

follows similarly. The product of the matrices $(\tilde{A}_i^j)$ and (A_j^k) is the unit matrix and thus these matrices are mutually inverse.

Exercise 2.7 Show that $x^i = \tilde{x}^k \tilde{A}_k^i$.

Formulas for the relations between reciprocal bases can be obtained as follows. We begin with the obvious identities

$$\mathbf{e}^j(\mathbf{e}_j \cdot \mathbf{x}) = \mathbf{x}, \qquad \tilde{\mathbf{e}}^j(\tilde{\mathbf{e}}_j \cdot \mathbf{x}) = \mathbf{x}.$$

Putting $\mathbf{x} = \tilde{\mathbf{e}}^i$ in the first of these gives

$$\tilde{\mathbf{e}}^i = A_j^i \mathbf{e}^j,$$

while the second identity with $\mathbf{x} = \mathbf{e}^i$ yields

$$\mathbf{e}^i = \tilde{A}_j^i \tilde{\mathbf{e}}^j.$$

From these follow the transformation formulas

$$\tilde{x}_i = x_k \tilde{A}_i^k, \qquad x_i = \tilde{x}_k A_i^k.$$

2.5 Covariant and Contravariant Components

We have seen that if the basis vectors are transformed according to the relation

$$\mathbf{e}_i = A_i^j \tilde{\mathbf{e}}_j,$$

then the components x_i of a vector $\mathbf{x}$ must transform according to

$$x_i = A_i^j \tilde{x}_j.$$

The similarity in form between these two relations results in the x_i being termed the *covariant components* of the vector $\mathbf{x}$. On the other hand, the transformation law

$$x^i = \tilde{A}_j^i \tilde{x}^j$$

shows that the x^i transform like the $\mathbf{e}^i$. For this reason the x^i are termed the *contravariant components* of $\mathbf{x}$. We shall find a further use for this nomenclature in Chapter 3.

Quick summary

If frame transformations are considered then $\mathbf{x}$ has the various expressions

$$\mathbf{x} = x^i \mathbf{e}_i = x_i \mathbf{e}^i = \tilde{x}^i \tilde{\mathbf{e}}_i = \tilde{x}_i \tilde{\mathbf{e}}^i$$

and the transformation laws

$$\tilde{x}^i = A^i_j x^j, \qquad \tilde{x}_i = \tilde{A}^j_i x_j, \qquad x^i = \tilde{A}^i_j \tilde{x}^j, \qquad x_i = A^j_i \tilde{x}_j,$$

apply. The x^i are termed contravariant components of $\mathbf{x}$, while the x_i are termed covariant components. The transformation laws are particularly simple when the frame is changed to the dual frame. Then

$$x_i = g_{ji} x^j, \qquad x^i = g^{ij} x_j,$$

where

$$g_{ij} = \mathbf{e}_i \cdot \mathbf{e}_j, \qquad g^{ij} = \mathbf{e}^i \cdot \mathbf{e}^j,$$

are components of the metric tensor.

2.6 The Cross Product in Index Notation

In mechanics a major role is played by the quantity called torque. This quantity is introduced in elementary physics as the product of a force magnitude and a length ("force times moment arm"), along with some rules for algebraic sign to account for the sense of rotation that the force would encourage when applied to a physical body. In more advanced discussions in which three-dimensional problems are considered, torque is regarded as a vectorial quantity. If a force $\mathbf{f}$ acts at a point which is located relative to an origin O by position vector $\mathbf{r}$, then the associated torque $\mathbf{t}$ about O is normal to the plane of the vectors $\mathbf{r}$ and $\mathbf{f}$. Of the two possible unit normals, $\mathbf{t}$ is conventionally (but arbitrarily) associated with the vector $\hat{\mathbf{n}}$ given by the familiar *right-hand rule*: if the forefinger of the right hand is directed along $\mathbf{r}$ and the middle finger is directed along $\mathbf{f}$, then the thumb indicates the direction of $\hat{\mathbf{n}}$ and hence the direction of $\mathbf{t}$. The magnitude of $\mathbf{t}$ equals $|\mathbf{f}||\mathbf{r}| \sin\theta$, where θ is the smaller angle between $\mathbf{f}$ and $\mathbf{r}$. These rules are all encapsulated in the brief symbolism

$$\mathbf{t} = \mathbf{r} \times \mathbf{f}.$$

The definition of torque can be taken as a model for a more general operation between vectors: the *cross product*. If $\mathbf{a}$ and $\mathbf{b}$ are any two vectors, we define

$$\mathbf{a} \times \mathbf{b} = \hat{\mathbf{n}}|\mathbf{a}||\mathbf{b}| \sin \theta$$

where $\hat{\mathbf{n}}$ and θ are defined as in the case of torque above. Like any other vector, $\mathbf{c} = \mathbf{a} \times \mathbf{b}$ can be expanded in terms of a basis; we choose the reciprocal basis $\mathbf{e}^i$ and write

$$\mathbf{c} = c_i \mathbf{e}^i.$$

Because the magnitudes of $\mathbf{a}$ and $\mathbf{b}$ enter into $\mathbf{a} \times \mathbf{b}$ in multiplicative fashion, we are prompted to seek c_i in the form

$$c_i = \epsilon_{ijk} a^j b^k. \tag{2.12}$$

Here the ϵ's are formal coefficients. Let us find them. We write

$$\mathbf{a} = a^j \mathbf{e}_j, \qquad \mathbf{b} = b^k \mathbf{e}_k,$$

and employ the well-known distributive property

$$(\mathbf{u} + \mathbf{v}) \times \mathbf{w} \equiv \mathbf{u} \times \mathbf{w} + \mathbf{v} \times \mathbf{w}$$

to obtain

$$\mathbf{c} = a^j \mathbf{e}_j \times b^k \mathbf{e}_k = a^j b^k (\mathbf{e}_j \times \mathbf{e}_k).$$

Then

$$\mathbf{c} \cdot \mathbf{e}_i = c_m \mathbf{e}^m \cdot \mathbf{e}_i = c_i = a^j b^k [(\mathbf{e}_j \times \mathbf{e}_k) \cdot \mathbf{e}_i]$$

and comparison with (2.12) shows that

$$\epsilon_{ijk} = (\mathbf{e}_j \times \mathbf{e}_k) \cdot \mathbf{e}_i.$$

Now the value of $(\mathbf{e}_j \times \mathbf{e}_k) \cdot \mathbf{e}_i$ depends on the values of the indices i, j, k. Here it is convenient to introduce the idea of a permutation of the ordered triple $(1, 2, 3)$. A permutation of $(1, 2, 3)$ is called *even* if it can be brought about by performing any even number of interchanges of pairs of these numbers; a permutation is *odd* if it results from performing any odd number of interchanges. We saw before that $(\mathbf{e}_j \times \mathbf{e}_k) \cdot \mathbf{e}_i$ is the volume V of the frame parallelepiped if i, j, k are distinct and the ordered triple (i, j, k) is an even permutation of $(1, 2, 3)$. If i, j, k are distinct and the ordered triple

(i,j,k) is an odd permutation of $(1,2,3)$, we obtain $-V$. If any two of the numbers i,j,k are equal we obtain zero. Hence

$$\epsilon_{ijk} = \begin{cases} +V, & (i,j,k) \text{ an even permutation of } (1,2,3), \\ -V, & (i,j,k) \text{ an odd permutation of } (1,2,3), \\ 0, & \text{two or more indices equal.} \end{cases}$$

Moreover, it can be shown (Exercise 2.4) that

$$V^2 = g$$

where g is the determinant of the matrix formed from the elements $g_{ij} = \mathbf{e}_i \cdot \mathbf{e}_j$ of the metric tensor. Note that $V = 1$ for a Cartesian frame.

The "permutation symbol" ϵ_{ijk} is useful in writing various formulas. For example, the determinant of a matrix $A = (a_{ij})$ can be expressed succinctly as

$$\det A = \epsilon_{ijk} a_{1i} a_{2j} a_{3k}.$$

Much more than a mere notational device however, ϵ_{ijk} represents a tensor (the so-called *Levi–Civita tensor*). We shall discuss this further in Chapter 3.

Exercise 2.8 The contravariant components of $\mathbf{c} = \mathbf{a} \times \mathbf{b}$ can be expressed as $c^i = \epsilon^{ijk} a_j b_k$ for suitable coefficients ϵ^{ijk}. Use the technique of this section to find the ϵ^{ijk}. Then establish the identity

$$\epsilon_{ijk}\epsilon^{pqr} = \begin{vmatrix} \delta_i^p & \delta_i^q & \delta_i^r \\ \delta_j^p & \delta_j^q & \delta_j^r \\ \delta_k^p & \delta_k^q & \delta_k^r \end{vmatrix}$$

and use it to show that

$$\epsilon_{ijk}\epsilon^{pqk} = \delta_i^p \delta_j^q - \delta_i^q \delta_j^p.$$

Use this in turn to prove that

$$\mathbf{a} \times (\mathbf{b} \times \mathbf{c}) = \mathbf{b}(\mathbf{a} \cdot \mathbf{c}) - \mathbf{c}(\mathbf{a} \cdot \mathbf{b}) \tag{2.13}$$

for any vectors $\mathbf{a}, \mathbf{b}, \mathbf{c}$.

Exercise 2.9 Establish Lagrange's identity

$$(\mathbf{a} \times \mathbf{b}) \cdot (\mathbf{c} \times \mathbf{d}) = (\mathbf{a} \cdot \mathbf{c})(\mathbf{b} \cdot \mathbf{d}) - (\mathbf{a} \cdot \mathbf{d})(\mathbf{b} \cdot \mathbf{c}).$$

2.7 Closing Remarks

We close by repeating something we said in Chapter 1: *a vector is an objective entity.* In elementary mathematics we learn to think of a vector as an ordered triple of components. There is, of course, no harm in this if we keep in mind a certain Cartesian frame. But if we fix those components then in any other frame the vector is determined uniquely. Absolutely uniquely! So a vector is something objective, but as soon as we specify its components in one frame we can find them in any other frame by the use of certain rules.

We choose to emphasize this here because the situation is exactly the same with tensors. A tensor is an objective entity, and fixing its components relative to one frame, we determine the tensor uniquely — even though its components relative to other frames will in general be different.

Chapter 3

Tensors

3.1 Dyadic Quantities and Tensors

We have met sets of quantities like g^{ij} or g_{ij}. Such a table of $3 \times 3 = 9$ coefficients could be considered as a vector in a nine-dimensional space, but we must reject this idea for an important reason: if we change the frame vectors and calculate the relations between the new and old components, the results differ in form from those that apply to vector components. The components of the metric tensor transform according to certain rules, however, and it is found that these transformation rules also apply to various quantities encountered in physical science. We indicated in Chapter 1 that these quantities, represented by 3×3 matrices, form a class of objects known as tensors of the second rank. Our plan is to present the relevant theory in a way that parallels the vector presentation of Chapter 2.

We begin to realize this program with the introduction of the *dyad* (or *tensor product*) of two vectors $\mathbf{a}$ and $\mathbf{b}$, denoted $\mathbf{a} \otimes \mathbf{b}$. We assume that the tensor product satisfies the usual properties of a product

$$
\begin{aligned}
(\lambda \mathbf{a}) \otimes \mathbf{b} &= \mathbf{a} \otimes (\lambda \mathbf{b}) = \lambda (\mathbf{a} \otimes \mathbf{b}), \\
(\mathbf{a} + \mathbf{b}) \otimes \mathbf{c} &= \mathbf{a} \otimes \mathbf{c} + \mathbf{b} \otimes \mathbf{c}, \\
\mathbf{a} \otimes (\mathbf{b} + \mathbf{c}) &= \mathbf{a} \otimes \mathbf{b} + \mathbf{a} \otimes \mathbf{c},
\end{aligned}
\tag{3.1}
$$

where λ is an arbitrary real number. From now on, however, we shall write out the dyad without the $\otimes$ symbol: $\mathbf{ab} = \mathbf{a} \otimes \mathbf{b}$.

Let us once again consider the space of three-dimensional vectors with the frame $\mathbf{e}_i$. Using the expansion of the vectors in the basis vectors and the properties (3.1), we represent the dyad $\mathbf{ab}$ as

$$\mathbf{ab} = a^i \mathbf{e}_i b^j \mathbf{e}_j = a^i b^j \mathbf{e}_i \mathbf{e}_j .$$

This introduces exactly nine different dyads $\mathbf{e}_i\mathbf{e}_j$. We now consider a linear space whose basis is this set of nine dyads, and call it the space of tensors of the second rank. The numerical coefficients of the dyads are called the components of the tensor. Thus an element of this space, a tensor $\mathbf{A}$, has the representation

$$\mathbf{A} = a^{ij}\mathbf{e}_i\mathbf{e}_j.$$

To maintain the property of objectivity of the elements of this space, we require that upon transformation of the frame the components of $\mathbf{A}$ transform correspondingly. Note that we have introduced superscript indices for the components of $\mathbf{A}$. This was done in keeping with the development of Chapter 2.

In preparation for the next section let us introduce the dot product of a dyad $\mathbf{ab}$ by a vector $\mathbf{c}$:

$$\mathbf{ab} \cdot \mathbf{c} = (\mathbf{b} \cdot \mathbf{c})\mathbf{a}. \tag{3.2}$$

So the result is a vector co-oriented with $\mathbf{a}$. Analogously we can introduce the dot product from the left:

$$\mathbf{c} \cdot (\mathbf{ab}) = (\mathbf{c} \cdot \mathbf{a})\mathbf{b}. \tag{3.3}$$

Exercise 3.1 (a) A dyad of the form $\mathbf{ee}$, where $\mathbf{e}$ is a unit vector, is sometimes called a *projection dyad.* Explain. (b) Write down matrices for the dyads $\mathbf{i}_1\mathbf{i}_1$, $\mathbf{i}_2\mathbf{i}_2$, and $\mathbf{i}_3\mathbf{i}_1$.

3.2 Tensors from an Operator Viewpoint

An alternative to viewing a second rank tensor as a weighted sum of dyads is to view the tensor as an *operator.* From this standpoint a tensor $\mathbf{A}$ is considered to map a vector $\mathbf{x}$ into a vector $\mathbf{y}$ according to the equation

$$\mathbf{y} = \mathbf{A} \cdot \mathbf{x}.$$

Conversely, a given linear relation between $\mathbf{x}$ and $\mathbf{y}$ will define the operator $\mathbf{A}$ uniquely. Thus if we have $\mathbf{A} \cdot \mathbf{x} = \mathbf{B} \cdot \mathbf{x}$ for all $\mathbf{x}$, then we have $\mathbf{A} = \mathbf{B}$. Let us show that the components are really uniquely defined in any basis by the equality $\mathbf{y} = \mathbf{A} \cdot \mathbf{x}$. $\mathbf{A}$ is represented by the expression $a^{ij}\mathbf{e}_i\mathbf{e}_j$ in some basis $\mathbf{e}_i$. It is clear that the operation $\mathbf{A} \cdot \mathbf{x}$ is linear in $\mathbf{x}$, and thus we define $\mathbf{A}$ uniquely if we specify its action on all three vectors of a basis.

Taking $\mathbf{x} = \mathbf{e}^k$, the corresponding $\mathbf{y}$ is

$$\mathbf{A} \cdot \mathbf{x} = a^{ij}\mathbf{e}_i\mathbf{e}_j \cdot \mathbf{e}^k = a^{ik}\mathbf{e}_i.$$

Dot multiplying this by $\mathbf{e}^l$ we get

$$a^{lk} = \mathbf{e}^l \cdot \mathbf{A} \cdot \mathbf{e}^k.$$

In this way we can find the components of a tensor $\mathbf{A}$ in any basis:

$$a^{ij} = \mathbf{e}^i \cdot \mathbf{A} \cdot \mathbf{e}^j, \quad a_{ij} = \mathbf{e}_i \cdot \mathbf{A} \cdot \mathbf{e}_j, \quad a^{i}_{\cdot j} = \mathbf{e}^i \cdot \mathbf{A} \cdot \mathbf{e}_j, \quad a_i^{\cdot j} = \mathbf{e}_i \cdot \mathbf{A} \cdot \mathbf{e}^j.$$

Note that in "mixed components" we position the indices in such a way that their association with the various dyads remains clear.

Analyzing the above reasoning, we can find that we have proved the so-called *quotient law* for tensors of the second rank. If $\mathbf{y}$ is a given vector and there is a linear transformation from $\mathbf{x}$ to $\mathbf{y}$ for an *arbitrary* vector $\mathbf{x}$, then the linear transformation is a tensor and we can write $\mathbf{y} = \mathbf{A} \cdot \mathbf{x}$. This statement is sometimes useful in establishing the tensorial character of a set of scalar quantities (i.e., the components of $\mathbf{A}$).

We may also define common algebraic operations from the operator viewpoint. Given tensors $\mathbf{A}$ and $\mathbf{B}$, the *sum* is the tensor $\mathbf{A} + \mathbf{B}$ uniquely defined by the requirement that

$$(\mathbf{A} + \mathbf{B}) \cdot \mathbf{x} = \mathbf{A} \cdot \mathbf{x} + \mathbf{B} \cdot \mathbf{x}$$

for all $\mathbf{x}$. If c is a scalar, $c\mathbf{A}$ is defined by the requirement that

$$(c\mathbf{A}) \cdot \mathbf{x} = c(\mathbf{A} \cdot \mathbf{x})$$

for all $\mathbf{x}$. In particular, any product of the form $0\mathbf{A}$ gives a *zero tensor* denoted $\mathbf{0}$. The dot product $\mathbf{A} \cdot \mathbf{B}$ is regarded as the composition of the operators $\mathbf{B}$ and $\mathbf{A}$:

$$(\mathbf{A} \cdot \mathbf{B}) \cdot \mathbf{x} \equiv \mathbf{A} \cdot (\mathbf{B} \cdot \mathbf{x}).$$

The dot product $\mathbf{y} \cdot \mathbf{A}$, called *pre-multiplication* of $\mathbf{A}$ by a vector $\mathbf{y}$, is defined by the requirement that

$$(\mathbf{y} \cdot \mathbf{A}) \cdot \mathbf{x} = \mathbf{y} \cdot (\mathbf{A} \cdot \mathbf{x})$$

for all vectors $\mathbf{x}$.

A simple but important tensor is the *unit tensor* denoted by $\mathbf{E}$ and defined by the requirement that for any $\mathbf{x}$

$$\mathbf{E} \cdot \mathbf{x} = \mathbf{x} \cdot \mathbf{E} = \mathbf{x}. \tag{3.4}$$

It is evident that in any Cartesian frame $\mathbf{i}_i$ we must have

$$\mathbf{E} = \sum_{i=1}^{3} \mathbf{i}_i \mathbf{i}_i. \tag{3.5}$$

In any frame we have

$$\mathbf{E} = \mathbf{e}^i \mathbf{e}_i = \mathbf{e}_j \mathbf{e}^j \tag{3.6}$$

for the mixed components. Consequently, the raising and lowering of indices gives

$$\mathbf{E} = g_{ij} \mathbf{e}^i \mathbf{e}^j = g^{ij} \mathbf{e}_i \mathbf{e}_j \tag{3.7}$$

in non-mixed components. We see that the role of the unit tensor belongs to the metric tensor! Throughout our discussion of second rank tensors we shall emphasize the close analogy between tensor theory and matrix theory. Equations (3.5) and (3.6) show that the matrix representation of $\mathbf{E}$ in either Cartesian or mixed components is the 3×3 identity matrix

$$\begin{pmatrix} 1 & 0 & 0 \\ 0 & 1 & 0 \\ 0 & 0 & 1 \end{pmatrix}.$$

This does not hold for the non-mixed components of (3.7).

Exercise 3.2 Use (3.4) along with (2.4) and (2.5) to show that the various components of $\mathbf{E}$ are given by

$$e^{ij} = g^{ij}, \qquad e_{ij} = g_{ij}, \qquad e^{i}_{\cdot j} = \delta^i_j, \qquad e_j^{\cdot i} = \delta^i_j.$$

Hence establish (3.6).

Our consideration of $\mathbf{A}$ as an operator leads us to introduce the notion of an *inverse tensor*: if

$$\mathbf{A} \cdot \mathbf{A}^{-1} = \mathbf{E},$$

then $\mathbf{A}^{-1}$ is called the inverse of $\mathbf{A}$. The inverse of a tensor is also a tensor. An important special case occurs when the matrix of the tensor is a *diagonal matrix.* If in a Cartesian frame $\mathbf{i}_i$ we have

$$\mathbf{A} = \sum_{i=1}^{3} \lambda_i \mathbf{i}_i \mathbf{i}_i$$

then the corresponding matrix representation is

$$\begin{pmatrix} \lambda_1 & 0 & 0 \\ 0 & \lambda_2 & 0 \\ 0 & 0 & \lambda_3 \end{pmatrix}.$$

If we take

$$\mathbf{B} = \sum_{j=1}^{3} \lambda_j^{-1} \mathbf{i}_j \mathbf{i}_j$$

to which there corresponds the matrix

$$\begin{pmatrix} \lambda_1^{-1} & 0 & 0 \\ 0 & \lambda_2^{-1} & 0 \\ 0 & 0 & \lambda_3^{-1} \end{pmatrix}$$

and form the dot product $\mathbf{A} \cdot \mathbf{B}$, we get

$$\begin{aligned} \mathbf{A} \cdot \mathbf{B} &= \sum_{i=1}^{3} \lambda_i \mathbf{i}_i \mathbf{i}_i \cdot \sum_{j=1}^{3} \lambda_j^{-1} \mathbf{i}_j \mathbf{i}_j = \sum_{i=1}^{3} \lambda_i \mathbf{i}_i \sum_{j=1}^{3} \lambda_j^{-1} (\mathbf{i}_i \cdot \mathbf{i}_j) \mathbf{i}_j \\ &= \sum_{i=1}^{3} \lambda_i \mathbf{i}_i \lambda_i^{-1} \mathbf{i}_i = \sum_{i=1}^{3} \mathbf{i}_i \mathbf{i}_i = \mathbf{E}. \end{aligned}$$

This means that $\mathbf{B} = \mathbf{A}^{-1}$. Correspondingly,

$$\begin{pmatrix} \lambda_1 & 0 & 0 \\ 0 & \lambda_2 & 0 \\ 0 & 0 & \lambda_3 \end{pmatrix} \begin{pmatrix} \lambda_1^{-1} & 0 & 0 \\ 0 & \lambda_2^{-1} & 0 \\ 0 & 0 & \lambda_3^{-1} \end{pmatrix} = \begin{pmatrix} 1 & 0 & 0 \\ 0 & 1 & 0 \\ 0 & 0 & 1 \end{pmatrix}.$$

Exercise 3.3 Establish the formula

$$(\mathbf{A} \cdot \mathbf{B})^{-1} = \mathbf{B}^{-1} \cdot \mathbf{A}^{-1}$$

for invertible tensors $\mathbf{A}$, $\mathbf{B}$.

A second rank tensor $\mathbf{A}$ is said to be *singular* if $\mathbf{A} \cdot \mathbf{x} = 0$ for some $\mathbf{x} \neq 0$. Hence $\mathbf{A}$ is called *nonsingular* if $\mathbf{A} \cdot \mathbf{x} = 0$ only when $\mathbf{x} = 0$. Recall that a matrix A is said to be nonsingular if and only if $\det A \neq 0$. The connection between the uses of this terminology in the two areas is as follows. If we take a mixed representation of the tensor $\mathbf{A}$, the equation $\mathbf{A} \cdot \mathbf{x} = 0$ yields a set of simultaneous equations in the components of $\mathbf{x}$; these equations have a nontrivial solution if and only if the determinant of the coefficient matrix (i.e., the matrix representing $\mathbf{A}$) is zero. Moreover, taking any other representation of $\mathbf{A}$ and a dual representation of the vector, we again arrive the same conclusion regarding the determinant. This brings the use of the term "singular" to the tensor $\mathbf{A}$. By definition the determinant of a second rank tensor $\mathbf{A}$, denoted $\det \mathbf{A}$, is the determinant of the matrix of its mixed components:

$$\det \mathbf{A} = \left| a_{i}^{\cdot j} \right| = \left| a^{k}_{\cdot m} \right| = \frac{1}{g} \left| a_{st} \right| = g \left| a^{pq} \right| .$$

The first equality is the definition as stated above; the other equalities are left for the reader to establish. Various other formulas such as

$$\det \mathbf{A} = \frac{1}{6} \epsilon_{ijk} \epsilon^{mnp} a_{m}^{\cdot i} a_{n}^{\cdot j} a_{p}^{\cdot k}$$

can be established for the determinant.

We close this section with an important remark. We can derive all desired properties of a tensor, and perform actions with the tensor, in any coordinate frame. Convenience will often dictate the use of Cartesian frames. But if we obtain an equation or expression through the use of a Cartesian frame and can subsequently represent this result in non-coordinate form, then we have provided rigorous justification of the latter. As we have said before, tensors are objective entities and ultimately all results pertaining to them must be frame independent.

3.3 Dyadic Components Under Transformation

The standpoint for deriving the transformation rules is that in any basis a tensor is the same element of some space, and only (3.1) and the rules we derived for vectors can govern the rules for transforming the components of a tensor. Let us begin with the transformation of the components when we go to the reciprocal basis. We set

$$a_{ij} \mathbf{e}^{i} \mathbf{e}^{j} = a^{ij} \mathbf{e}_{i} \mathbf{e}_{j}$$

and take dot products as in (3.2) and (3.3):

$$\mathbf{e}_{k} \cdot a_{ij} \mathbf{e}^{i} \mathbf{e}^{j} \cdot \mathbf{e}_{m} = \mathbf{e}_{k} \cdot a^{ij} \mathbf{e}_{i} \mathbf{e}_{j} \cdot \mathbf{e}_{m}.$$

This gives

$$a_{ij}(\mathbf{e}_k \cdot \mathbf{e}^i)(\mathbf{e}^j \cdot \mathbf{e}_m) = a^{ij}(\mathbf{e}_k \cdot \mathbf{e}_i)(\mathbf{e}_j \cdot \mathbf{e}_m),$$

hence

$$a_{km} = a^{ij} g_{ki} g_{jm}.$$

We see that the components of the metric tensor are encountered in this transformation.

Now we can construct the formulas for transforming the tensor components when the change of basis takes the general form

$$\mathbf{e}_i = A_i^j \tilde{\mathbf{e}}_j.$$

From

$$\tilde{a}^{ij} \tilde{\mathbf{e}}_i \tilde{\mathbf{e}}_j = a^{km} \mathbf{e}_k \mathbf{e}_m = a^{km} A_k^p \tilde{\mathbf{e}}_p A_m^q \tilde{\mathbf{e}}_q$$

we obtain

$$\tilde{a}^{ij} = a^{km} A_k^i A_m^j. \tag{3.8}$$

Similarly, the inverse transformation

$$\tilde{\mathbf{e}}_i = \tilde{A}_i^j \mathbf{e}_j$$

leads to

$$a^{ij} = \tilde{a}^{km} \tilde{A}_k^i \tilde{A}_m^j. \tag{3.9}$$

Equations (3.8) and (3.9) together imply that

$$A_j^k \tilde{A}_k^i = \delta_j^i.$$

Various expressions for $\mathbf{A}$,

$$\begin{aligned} \mathbf{A} &= a^{ij}\mathbf{e}_i\mathbf{e}_j = a_{kl}\mathbf{e}^k\mathbf{e}^l = a_i^{\cdot j}\mathbf{e}^i\mathbf{e}_j = a^k_{\cdot l}\mathbf{e}_k\mathbf{e}^l \\ &= \tilde{a}^{ij}\tilde{\mathbf{e}}_i\tilde{\mathbf{e}}_j = \tilde{a}_{kl}\tilde{\mathbf{e}}^k\tilde{\mathbf{e}}^l = \tilde{a}_i^{\cdot j}\tilde{\mathbf{e}}^i\tilde{\mathbf{e}}_j = \tilde{a}^k_{\cdot l}\tilde{\mathbf{e}}_k\tilde{\mathbf{e}}^l, \end{aligned}$$

lead to other transformation formulas such as

$$\tilde{a}_{ij} = \tilde{A}_i^k \tilde{A}_j^l a_{kl}, \qquad a_{ij} = A_i^k A_j^l \tilde{a}_{kl},$$

and

$$\tilde{a}_i^{\cdot j} = \tilde{A}_i^k A_l^j a_k^{\cdot l}, \qquad \tilde{a}^i_{\cdot j} = A_k^i \tilde{A}_j^l a^k_{\cdot l}.$$

Remembering the terminology of § 2.5, we see why the a_{ij} are called the covariant components of **A** while the a^{ij} are called the contravariant components. The components $a^{i}_{\cdot j}$ and $a_{i}^{\cdot j}$ are called mixed components.

Quick summary

We have

$$\begin{aligned}\mathbf{A} &= \tilde{a}^{ij}\tilde{\mathbf{e}}_i\tilde{\mathbf{e}}_j = \tilde{a}_{kl}\tilde{\mathbf{e}}^k\tilde{\mathbf{e}}^l = \tilde{a}_i^{\cdot j}\tilde{\mathbf{e}}^i\tilde{\mathbf{e}}_j = \tilde{a}^{k}_{\cdot l}\tilde{\mathbf{e}}_k\tilde{\mathbf{e}}^l \\ &= a^{ij}\mathbf{e}_i\mathbf{e}_j = a_{kl}\mathbf{e}^k\mathbf{e}^l = a_i^{\cdot j}\mathbf{e}^i\mathbf{e}_j = a^{k}_{\cdot l}\mathbf{e}_k\mathbf{e}^l\end{aligned}$$

where

$$\begin{aligned}
\tilde{a}^{ij} &= A^i_k A^j_l a^{kl}, & a^{ij} &= \tilde{A}^i_k \tilde{A}^j_l \tilde{a}^{kl}, \\
\tilde{a}_{ij} &= \tilde{A}^k_i \tilde{A}^l_j a_{kl}, & a_{ij} &= A^k_i A^l_j \tilde{a}_{kl}, \\
\tilde{a}^{i}_{\cdot j} &= A^i_k \tilde{A}^l_j a^{k}_{\cdot l}, & a^{i}_{\cdot j} &= \tilde{A}^i_k A^l_j \tilde{a}^{k}_{\cdot l}, \\
\tilde{a}_i^{\cdot j} &= \tilde{A}^k_i A^j_l a_k^{\cdot l}, & a_i^{\cdot j} &= A^k_i \tilde{A}^j_l \tilde{a}_k^{\cdot l}.
\end{aligned}$$

Exercise 3.4 (a) Express the transformation law $\tilde{b}^{ij} = A^i_k A^j_m b^{km}$ in matrix notation. (b) Repeat for a transformation law of the form $\tilde{b}_{ij} = A^k_i A^m_j b_{km}$.

Exercise 3.5 Our A^j_i values give the transformation from one basis to another; they define a transformation of the space that is an operator, and hence a tensor of the second rank. Write out the tensor for which the A^j_i are components. Repeat for the inverse transformation.

3.4 More Dyadic Operations

The dot product of two dyads **ab** and **cd** is defined by

$$\mathbf{ab} \cdot \mathbf{cd} = (\mathbf{b} \cdot \mathbf{c})\mathbf{ad}.$$

The result is again a dyad, with a coefficient $\mathbf{b} \cdot \mathbf{c}$. Extensions of this and the formulas (3.2) and (3.3) to operations with sums of dyads and vectors (using (3.1) and the vectorial rules) gives us a number of rules which the dot product obeys. Let **A** and **B** be dyads, **a** and **b** be vectors, and λ and

μ be any real numbers. Then

$$\mathbf{A} \cdot (\lambda \mathbf{a} + \mu \mathbf{b}) = \lambda \mathbf{A} \cdot \mathbf{a} + \mu \mathbf{A} \cdot \mathbf{b},$$
$$(\lambda \mathbf{A} + \mu \mathbf{B}) \cdot \mathbf{a} = \lambda \mathbf{A} \cdot \mathbf{a} + \mu \mathbf{B} \cdot \mathbf{a}.$$

Similar identities hold for dot products taken in the opposite orders. These results show that linearity may be assumed in working with these operations.

Now let us pursue a close analogy that exists between the dot product and matrix multiplication. We begin with the simple case of a Cartesian frame. We take a dyad $\mathbf{A}$ and a vector $\mathbf{b}$ and express these relative to a basis $\mathbf{i}_k$:

$$\mathbf{A} = a^{km}\mathbf{i}_k\mathbf{i}_m, \qquad \mathbf{b} = b^j\mathbf{i}_j.$$

Denoting

$$\mathbf{c} = \mathbf{A} \cdot \mathbf{b} \tag{3.10}$$

we have $c^k\mathbf{i}_k = a^{km}\mathbf{i}_k\mathbf{i}_m \cdot b^j\mathbf{i}_j$ so that

$$c^k = \sum_{j=1}^{3} a^{kj} b^j.$$

(We have inserted the summation symbol because j stands in the upper position twice and the summation convention would not apply.) Written out, this is the system of three equations

$$c^1 = a^{11}b^1 + a^{12}b^2 + a^{13}b^3,$$
$$c^2 = a^{21}b^1 + a^{22}b^2 + a^{23}b^3,$$
$$c^3 = a^{31}b^1 + a^{32}b^2 + a^{33}b^3,$$

or

$$\begin{pmatrix} c^1 \\ c^2 \\ c^3 \end{pmatrix} = \begin{pmatrix} a^{11} & a^{12} & a^{13} \\ a^{21} & a^{22} & a^{23} \\ a^{31} & a^{32} & a^{33} \end{pmatrix} \begin{pmatrix} b^1 \\ b^2 \\ b^3 \end{pmatrix}.$$

Here we have a matrix equation of the form

$$c = Ab \tag{3.11}$$

where c and b are column vectors and A is a 3×3 matrix. The analogy between the dot product and matrix multiplication is evident from (3.10) and (3.11). This analogy extends beyond the confines of Cartesian frames. Let us write, for example,

$$\mathbf{A} = a^{km}\mathbf{e}_k\mathbf{e}_m, \qquad \mathbf{b} = b_j\mathbf{e}^j.$$

This time $\mathbf{c} = \mathbf{A} \cdot \mathbf{b}$ gives $c^k\mathbf{e}_k = a^{km}\mathbf{e}_k\mathbf{e}_m \cdot b_j\mathbf{e}^j = a^{km}\mathbf{e}_k\delta^j_m b_j$, hence

$$c^k = a^{kj}b_j.$$

The corresponding matrix equation is, of course,

$$\begin{pmatrix} c^1 \\ c^2 \\ c^3 \end{pmatrix} = \begin{pmatrix} a^{11} & a^{12} & a^{13} \\ a^{21} & a^{22} & a^{23} \\ a^{31} & a^{32} & a^{33} \end{pmatrix} \begin{pmatrix} b_1 \\ b_2 \\ b_3 \end{pmatrix}.$$

With suitable understanding we could still write this as (3.11). Note what happens when we express both the dyad and the vector in terms of covariant components:

$$\mathbf{A} = a_{km}\mathbf{e}^k\mathbf{e}^m, \qquad \mathbf{b} = b_j\mathbf{e}^j.$$

We obtain

$$c_k = a_{km}g^{mj}b_j$$

and the metric tensor appears. The corresponding matrix form is

$$\begin{pmatrix} c_1 \\ c_2 \\ c_3 \end{pmatrix} = \begin{pmatrix} a_{11} & a_{12} & a_{13} \\ a_{21} & a_{22} & a_{23} \\ a_{31} & a_{32} & a_{33} \end{pmatrix} \begin{pmatrix} g^{11} & g^{12} & g^{13} \\ g^{21} & g^{22} & g^{23} \\ g^{31} & g^{32} & g^{33} \end{pmatrix} \begin{pmatrix} b_1 \\ b_2 \\ b_3 \end{pmatrix}.$$

Because the metric tensor can raise an index on a vector component, we may also write these equations in the forms

$$c_k = a_{km}b^m$$

and

$$\begin{pmatrix} c_1 \\ c_2 \\ c_3 \end{pmatrix} = \begin{pmatrix} a_{11} & a_{12} & a_{13} \\ a_{21} & a_{22} & a_{23} \\ a_{31} & a_{32} & a_{33} \end{pmatrix} \begin{pmatrix} b^1 \\ b^2 \\ b^3 \end{pmatrix}.$$

Let us now examine the dot product between two dyads. There are various possibilities for the components of the dyad

$$\mathbf{C} = \mathbf{A} \cdot \mathbf{B},$$

depending on how we choose to express **A** and **B**. If we use all contravariant components and write

$$\mathbf{A} = a^{km}\mathbf{e}_k\mathbf{e}_m, \qquad \mathbf{B} = b^{km}\mathbf{e}_k\mathbf{e}_m,$$

then $\mathbf{C} = c^{kn}\mathbf{e}_k\mathbf{e}_n$ where

$$c^{kn} = a^{km}g_{mj}b^{jn}.$$

Similarly, the use of all covariant components as in

$$\mathbf{A} = a_{km}\mathbf{e}^k\mathbf{e}^m, \qquad \mathbf{B} = b_{km}\mathbf{e}^k\mathbf{e}^m,$$

leads to $\mathbf{C} = c_{kn}\mathbf{e}^k\mathbf{e}^n$ where

$$c_{kn} = a_{km}g^{mj}b_{jn}.$$

Mixed components appear when we express

$$\mathbf{A} = a^{km}\mathbf{e}_k\mathbf{e}_m, \qquad \mathbf{B} = b_{km}\mathbf{e}^k\mathbf{e}^m.$$

Then

$$\mathbf{C} = \mathbf{A}\cdot\mathbf{B} = a^{km}\mathbf{e}_k\mathbf{e}_m \cdot b_{jn}\mathbf{e}^j\mathbf{e}^n = a^{km}\delta^j_m b_{jn}\mathbf{e}_k\mathbf{e}^n = a^{kj}b_{jn}\mathbf{e}_k\mathbf{e}^n.$$

Defining $c^k_{\cdot n} = a^{kj}b_{jn}$ we have $\mathbf{C} = c^k_{\cdot n}\mathbf{e}_k\mathbf{e}^n$. We leave the consideration of various other possibilities to the reader as

Exercise 3.6 (a) Discuss how the formulation $c_k^{\cdot n} = a_{kj}b^{jn}$ arises. (b) Show how all the forms above correspond to matrix multiplication. (c) What happens if mixed components are used on the right-hand sides to express **A** and **B**?

Another useful operation that can be performed between tensors is *double dot multiplication*. If **ab** and **cd** are dyads, we define

$$\mathbf{ab} \cdot\cdot\, \mathbf{cd} = (\mathbf{b}\cdot\mathbf{c})(\mathbf{a}\cdot\mathbf{d}).$$

That is, we first dot multiply the near standing vectors, then the remaining vectors, and thereby obtain a scalar as the result.

Exercise 3.7 (a) Calculate $\mathbf{A} \cdot\cdot\, \mathbf{E}$ if **A** is a tensor of rank two. How does this relate to the trace of the matrix that represents **A** in mixed components? (b) Let **A** and **B** be tensors of rank two. Write down several different component forms for the quantity $\mathbf{A} \cdot\cdot\, \mathbf{B}$.

3.5 Properties of Second Rank Tensors

Now we would like to consider in more detail those tensors that occur most frequently in applications: tensors of the second rank. First we recall that such a tensor is represented in dyadic form as

$$\mathbf{A} = a^{ij}\mathbf{e}_i\mathbf{e}_j.$$

Those who work in the applied sciences are probably more accustomed to the matrix representation

$$\begin{pmatrix} a^{11} & a^{12} & a^{13} \\ a^{21} & a^{22} & a^{23} \\ a^{31} & a^{32} & a^{33} \end{pmatrix}. \tag{3.12}$$

When we use the matrix form (3.12) the dyadic basis of the tensor remains implicit. Of course when we use a unique, say Cartesian, frame for the space of vectors, then it does not matter whether we show the dyads. The correspondence between the dyadic and matrix representations suggests that we can introduce many familiar ideas from the theory of matrices.

The tensor transpose

Let us begin with the notion of *transposition*. For a matrix $A = (a^{ij})$ the transposed matrix A^T is

$$A^T = (a^{ji}).$$

Similarly we introduce the transpose operation for the tensor $\mathbf{A}$:

$$\mathbf{A}^T = a^{ji}\mathbf{e}_i\mathbf{e}_j. \tag{3.13}$$

Let us note that this operation yields a new tensor, in each representation of which the corresponding indices appear in reverse order:

$$\mathbf{A}^T = a^{ji}\mathbf{e}_i\mathbf{e}_j = a_{ji}\mathbf{e}^i\mathbf{e}^j = a^{j}_{\cdot i}\mathbf{e}^i\mathbf{e}_j = a_j^{\cdot i}\mathbf{e}_i\mathbf{e}^j.$$

A useful relation that holds for any second rank tensor $\mathbf{A}$ and any vector $\mathbf{x}$ is

$$\mathbf{A} \cdot \mathbf{x} = \mathbf{x} \cdot \mathbf{A}^T. \tag{3.14}$$

This follows when we write $\mathbf{x} = x^k\mathbf{e}_k$ and note from (3.13) that $\mathbf{A}^T = a^{ij}\mathbf{e}_j\mathbf{e}_i$. Equation (3.14) is sometimes used to define the transpose. Also note that

$$(\mathbf{A}^T)^T = \mathbf{A}$$

for any second rank tensor $\mathbf{A}$.

Exercise 3.8 (a) Show that if $\mathbf{A}$ and $\mathbf{B}$ are tensors of the second rank, then $(\mathbf{A} \cdot \mathbf{B})^T = \mathbf{B}^T \cdot \mathbf{A}^T$. (b) Let $\mathbf{a}$ and $\mathbf{b}$ be vectors and $\mathbf{C}$ be a tensor of second rank. Show that $\mathbf{a} \cdot \mathbf{C}^T \cdot \mathbf{b} = \mathbf{b} \cdot \mathbf{C} \cdot \mathbf{a}$. (c) Show that if $\mathbf{A}$ is a nonsingular tensor of the second rank, then the components of the tensor $\mathbf{B} = \mathbf{A}^{-1}$ are given by the formulas

$$b_i^{\cdot j} = \frac{1}{2 \det \mathbf{A}} \epsilon_{ikl} \epsilon^{jmn} a_m^{\cdot k} a_n^{\cdot l}.$$

(d) Verify the following relations:

$$\det \mathbf{A}^{-1} = (\det \mathbf{A})^{-1}, \qquad (\mathbf{A} \cdot \mathbf{B})^{-1} = \mathbf{B}^{-1} \cdot \mathbf{A}^{-1},$$
$$(\mathbf{A}^T)^{-1} = (\mathbf{A}^{-1})^T, \qquad (\mathbf{A}^{-1})^{-1} = \mathbf{A}.$$

Tensors raised to powers

By analogy with matrix algebra we may raise a tensor to a positive integer power:

$$\mathbf{A}^2 = \mathbf{A} \cdot \mathbf{A}, \qquad \mathbf{A}^3 = \mathbf{A} \cdot \mathbf{A}^2, \qquad \mathbf{A}^4 = \mathbf{A} \cdot \mathbf{A}^3,$$

and so on. Negative integer powers are defined by raising $\mathbf{A}^{-1}$ to positive integer powers:

$$\mathbf{A}^{-2} = \mathbf{A}^{-1} \cdot \mathbf{A}^{-1}, \qquad \mathbf{A}^{-3} = \mathbf{A}^{-2} \cdot \mathbf{A}^{-1}, \qquad \mathbf{A}^{-4} = \mathbf{A}^{-3} \cdot \mathbf{A}^{-1},$$

and so on. These operations can be used to construct functions of tensors using Taylor expansions of elementary functions. For example, $e^x = 1 + x/1! + x^2/2! + x^3/3! + \cdots$. By this we can introduce the exponential of the tensor $\mathbf{A}$:

$$e^{\mathbf{A}} = \mathbf{E} + \frac{\mathbf{A}}{1!} + \frac{\mathbf{A}^2}{2!} + \frac{\mathbf{A}^3}{3!} + \cdots.$$

The issue of convergence of such series is approached in a manner similar to the absolute convergence of usual series, but with use of a norm of the tensor $\mathbf{A}$ (see § 4.11). We can introduce other functions similarly. This technique is used in the study of nonlinear elasticity, for example.

Symmetric and antisymmetric tensors

Among the class of all second rank tensors, an important role is played by the *symmetric* tensors. These include the strain and stress tensors of the theory of elasticity. The tensor of inertia is symmetric, as is the metric tensor. All these satisfy the relation

$$\mathbf{A} = \mathbf{A}^T.$$

It follows from (3.14) that

$$\mathbf{A} \cdot \mathbf{x} = \mathbf{x} \cdot \mathbf{A}$$

if $\mathbf{A}$ is symmetric. The reader will recall that the unit tensor $\mathbf{E}$ satisfies a relation of this form. A tensor $\mathbf{A}$ is said to be *antisymmetric* if

$$\mathbf{A} = -\mathbf{A}^T.$$

Exercise 3.9 Give the matrix forms corresponding to the cases of symmetric and antisymmetric tensors. How many components can be independently specified for a symmetric tensor? For an antisymmetric tensor?

Both symmetric and antisymmetric tensors arise naturally in the physical sciences. Their significance is also shown by the following

Theorem 3.1 *Any tensor of second rank can be decomposed as a sum of symmetric and antisymmetric tensors:*

$$\mathbf{A} = \mathbf{B} + \mathbf{C}$$

where $\mathbf{B} = \mathbf{B}^T$ *and* $\mathbf{C} = -\mathbf{C}^T$.

Proof. Take

$$\mathbf{B} = \frac{1}{2}\left(\mathbf{A} + \mathbf{A}^T\right), \qquad \mathbf{C} = \frac{1}{2}\left(\mathbf{A} - \mathbf{A}^T\right),$$

and check all the statements. □

The dyad $\mathbf{ab}$ can be decomposed into symmetric and antisymmetric parts as

$$\mathbf{ab} = \frac{1}{2}(\mathbf{ab} + \mathbf{ba}) + \frac{1}{2}(\mathbf{ab} - \mathbf{ba})$$

for example.

Exercise 3.10 Show that if $\mathbf{A}$ is symmetric and $\mathbf{B}$ is antisymmetric then $\mathbf{A} \cdot\cdot\, \mathbf{B} = 0$.

Exercise 3.11 Demonstrate that the quadratic form $\mathbf{x} \cdot \mathbf{A} \cdot \mathbf{x}$ does not change if the second rank tensor $\mathbf{A}$ is replaced by its symmetric part.

Given an antisymmetric tensor $\mathbf{C}$ written in a Cartesian frame as $\mathbf{C} = c^{ij}\mathbf{i}_i\mathbf{i}_j$, we can construct a vector $\boldsymbol{\omega} = \omega^k \mathbf{i}_k$ according to the formulas

$$\omega^1 = c^{32}, \qquad \omega^2 = c^{13}, \qquad \omega^3 = c^{21}.$$

It is easy to verify directly that

$$\mathbf{C} \cdot \mathbf{x} = \boldsymbol{\omega} \times \mathbf{x}, \qquad \mathbf{x} \cdot \mathbf{C} = \mathbf{x} \times \boldsymbol{\omega},$$

where $\mathbf{x}$ is an arbitrary vector. These formulas are written in non-coordinate form so they hold in any frame. The reader can derive the formulas for $\boldsymbol{\omega}$, which is called the conjugate vector, for an arbitrary frame.

Eigenvalues and eigenvectors

We now consider the question of which basis yields a tensor of simplest form. As the analogous question in matrix theory relates to eigenvalues and eigenvectors, we extend these notions to tensors. The pair $(\lambda, \mathbf{x})$, $\mathbf{x} \neq 0$, is called an *eigenpair* if the equality

$$\mathbf{A} \cdot \mathbf{x} = \lambda \mathbf{x} \tag{3.15}$$

holds. Hence $\mathbf{x}$ is an eigenvector of $\mathbf{A}$ if $\mathbf{A}$ operates on $\mathbf{x}$ to give a vector proportional to $\mathbf{x}$. Equation (3.15) may also be written in the form

$$(\mathbf{A} - \lambda \mathbf{E}) \cdot \mathbf{x} = 0.$$

Exercise 3.12 Find the eigenpairs of the dyad $\mathbf{ab}$. Now try to position an eigenvector on the left: $\mathbf{x} \cdot \mathbf{ab} = \lambda \mathbf{x}$. You should find that this differs from the previous eigenvector, and so it is sensible to introduce left and right eigenvectors. In the case of a symmetric tensor they coincide.

The eigenvalues of a second rank tensor $\mathbf{A}$ are found as solutions of the characteristic equation for $\mathbf{A}$, which is derived as follows. Using components (3.15) becomes

$$a^{ij}\mathbf{e}_i\mathbf{e}_j \cdot x^k \mathbf{e}_k = \lambda x^i \mathbf{e}_i$$

or

$$a^{ij} g_{jk} x^k \mathbf{e}_i = \lambda x^k \delta^i_k \mathbf{e}_i.$$

Writing this as

$$(a^{i}_{\cdot k} - \lambda \delta^{i}_{k}) x^{k} = 0,$$

we have a system of three simultaneous equations in the three variables x^k. A non-trivial solution exists if and only if the determinant of the coefficient matrix vanishes:

$$\begin{vmatrix} a^{1}_{\cdot 1} - \lambda & a^{1}_{\cdot 2} & a^{1}_{\cdot 3} \\ a^{2}_{\cdot 1} & a^{2}_{\cdot 2} - \lambda & a^{2}_{\cdot 3} \\ a^{3}_{\cdot 1} & a^{3}_{\cdot 2} & a^{3}_{\cdot 3} - \lambda \end{vmatrix} = 0.$$

This is the characteristic equation[1] for **A**. Writing it in the form

$$-\lambda^3 + I_1(\mathbf{A})\lambda^2 - I_2(\mathbf{A})\lambda + I_3(\mathbf{A}) = 0 \tag{3.16}$$

we note that it is cubic in λ, hence there are at most three distinct eigenvalues λ_1, λ_2, λ_3. The coefficients $I_1(\mathbf{A})$, $I_2(\mathbf{A})$, and $I_3(\mathbf{A})$ are called the *first, second, and third invariants* of **A**, and are expressed in terms of the eigenvalues by the *Viète formulas*

$$\begin{aligned} I_1(\mathbf{A}) &= \lambda_1 + \lambda_2 + \lambda_3, \\ I_2(\mathbf{A}) &= \lambda_1\lambda_2 + \lambda_1\lambda_3 + \lambda_2\lambda_3, \\ I_3(\mathbf{A}) &= \lambda_1\lambda_2\lambda_3. \end{aligned}$$

After representing the tensor in diagonal form it will be easy to see that $I_1(\mathbf{A})$ and $I_3(\mathbf{A})$ are, respectively, the trace and the determinant of the tensor **A**. In nonlinear elasticity the invariants play an important role, as they participate in the formulation of various constitutional laws.

Exercise 3.13 A tensor **A**, when referred to a certain Cartesian basis, has matrix

$$\begin{pmatrix} 1 & 0 & 1 \\ 2 & -1 & 0 \\ 0 & 1 & 2 \end{pmatrix}.$$

Find the first, second, and third principal invariants of **A**.

The most important second rank tensors are the symmetric tensors, and for these we may establish special properties. We begin by extending the definition of the vector dot product to cover the complex case. When **x** and

[1]Note that it is expressed in terms of mixed components of the tensor. However, it characterizes the properties of the tensor and has invariant properties since the eigenvalues of tensor do not depend on the coordinate frame in which they were obtained.

$\mathbf{y}$ are complex, a complex conjugate operation is taken on the components of $\mathbf{y}$ in performing the operation $\mathbf{x} \cdot \mathbf{y}$. We have, for instance,

$$\mathbf{x} \cdot \mathbf{y} = x^i \mathbf{e}_i \cdot \overline{y}^j \mathbf{e}_j = x^i \overline{y}_i.$$

Note that this reduces to our previous definition of the dot product if $\mathbf{x}$ and $\mathbf{y}$ are purely real.

Theorem 3.2 *The eigenvalues of a real symmetric tensor are real. Moreover, eigenvectors corresponding to distinct eigenvalues are orthogonal.*

Proof. Let $\mathbf{A}$ be a real symmetric tensor. If λ is an eigenvalue of $\mathbf{A}$, then we may dot $\mathbf{x}$ with both sides of (3.15) and write the result as

$$\lambda = \frac{\mathbf{x} \cdot (\mathbf{A} \cdot \mathbf{x})}{\mathbf{x} \cdot \mathbf{x}}$$

since $\mathbf{x} \neq 0$. Here $\mathbf{x} \cdot \mathbf{x} = |\mathbf{x}|^2$ is real. To see that $\mathbf{x} \cdot (\mathbf{A} \cdot \mathbf{x})$ is also real, we write

$$\begin{aligned} \overline{\mathbf{x} \cdot (\mathbf{A} \cdot \mathbf{x})} &= \overline{x^i \mathbf{e}_i \cdot a^{jk} \mathbf{e}_j \mathbf{e}_k \cdot \overline{x}^m \mathbf{e}_m} \\ &= \overline{x}^i \mathbf{e}_i \cdot a^{jk} \mathbf{e}_k \mathbf{e}_j \cdot x^m \mathbf{e}_m \\ &= x^m \mathbf{e}_m \cdot a^{jk} \mathbf{e}_j \mathbf{e}_k \cdot \overline{x}^i \mathbf{e}_i \\ &= \mathbf{x} \cdot (\mathbf{A} \cdot \mathbf{x}) \end{aligned}$$

(we used $a^{jk} = a^{kj}$). Because the eigenvalues λ are real, the components of eigenvectors satisfy a linear system of simultaneous equations having real coefficients, and so they are real as well.

To prove the second part of the theorem, suppose that $\mathbf{A} \cdot \mathbf{x}_1 = \lambda_1 \mathbf{x}_1$ and $\mathbf{A} \cdot \mathbf{x}_2 = \lambda_2 \mathbf{x}_2$ where $\lambda_2 \neq \lambda_1$. From these we obtain

$$\lambda_1 \mathbf{x}_1 \cdot \mathbf{x}_2 = (\mathbf{A} \cdot \mathbf{x}_1) \cdot \mathbf{x}_2, \qquad \lambda_2 \mathbf{x}_2 \cdot \mathbf{x}_1 = (\mathbf{A} \cdot \mathbf{x}_2) \cdot \mathbf{x}_1,$$

and subtraction gives

$$(\lambda_1 - \lambda_2) \mathbf{x}_1 \cdot \mathbf{x}_2 = \mathbf{x}_2 \cdot \mathbf{A} \cdot \mathbf{x}_1 - \mathbf{x}_1 \cdot \mathbf{A} \cdot \mathbf{x}_2 = 0.$$

Here we used the fact that the eigenvectors are real (to interchange the order of vectors in the dot products) and the fact that $\mathbf{A}$ is real and symmetric (Exercise 3.8). □

Exercise 3.14 Show that for any second rank tensor $\mathbf{A}$ (not necessarily symmetric), eigenvectors corresponding to distinct eigenvalues are linearly independent.

In solving the characteristic equation for λ, we may find that there are three distinct solutions or fewer than three. Note that if $\mathbf{x}$ is an eigenvector of $\mathbf{A}$ then so is $\alpha\mathbf{x}$ for any $\alpha \neq 0$. In other words, an eigenvector is determined up to a constant multiple. So when the eigenvalues are distinct we can compose a Cartesian frame from the orthonormal eigenvectors $\mathbf{x}_k$ and then express the tensor $\mathbf{A}$ in terms of its components a_{ij} as

$$\mathbf{A} = \sum a_{ij}\mathbf{x}_i\mathbf{x}_j.$$

Since the frame $\mathbf{x}_k$ is Cartesian the reciprocal basis is the same, and we may calculate the components of $\mathbf{A}$ from

$$a_{ij} = \mathbf{x}_i \cdot \mathbf{A} \cdot \mathbf{x}_j.$$

Since the $\mathbf{x}_i$ are also eigenvectors we can write

$$\mathbf{A} \cdot \mathbf{x}_j = \lambda_j \mathbf{x}_j,$$

and dot multiplication by $\mathbf{x}_i$ from the left gives

$$\mathbf{x}_i \cdot \mathbf{A} \cdot \mathbf{x}_j = \mathbf{x}_i \cdot \lambda_j \mathbf{x}_j = \lambda_j \delta_i^j.$$

Hence the coefficients of the dyads $\mathbf{x}_i\mathbf{x}_j$ in $\mathbf{A}$ are nonzero only for those coefficients that lie on the main diagonal of the matrix representation of $\mathbf{A}$; moreover, we see that these diagonal entries are the eigenvalues of $\mathbf{A}$. We can therefore write

$$\mathbf{A} = \sum_{i=1}^{3} \lambda_i \mathbf{x}_i \mathbf{x}_i.$$

This is called the *orthogonal representation* of $\mathbf{A}$. The eigenvectors composing the coordinate frame give us the *principal axes* of $\mathbf{A}$, and the process of referring the tensor to its principal axes is known as *diagonalization.*

When the characteristic equation of a tensor has fewer than three distinct solutions for λ, then one of these is said to be *degenerate.* If $\mathbf{A}$ is symmetric we may draw some relevant conclusions by representing $\mathbf{A}$ in a Cartesian basis and then applying facts from the theory of symmetric matrices. Corresponding to a multiple root of the characteristic equation we have a subspace of eigenvectors. If $\lambda_1 = \lambda_2$ is a double root, then the subspace is two-dimensional and we can select an orthonormal pair that is orthogonal to the third eigenvector (since λ_1 and λ_3 are distinct). Such an eigenvalue corresponding to two linearly independent eigenvectors is regarded as a *multiple eigenvalue* (multiplicity two). If $\lambda_1 = \lambda_2 = \lambda_3$ then any vector is an eigenvector, and choosing a Cartesian frame we would have an eigenvalue of multiplicity three. However, this case arises only when the tensor under consideration is proportional to the unit tensor $\mathbf{E}$. Such a tensor is called a *ball tensor.*

Exercise 3.15 Show directly that a second rank tensor (not necessarily symmetric) having three distinct eigenvalues cannot have more than one linearly independent eigenvector corresponding to each eigenvalue.

The Cayley–Hamilton theorem

The *Cayley–Hamilton theorem* states that every square matrix satisfies its own characteristic equation. For example the 2×2 matrix

$$A = \begin{pmatrix} a & b \\ c & d \end{pmatrix}$$

has characteristic equation

$$\begin{vmatrix} a-\lambda & b \\ c & d-\lambda \end{vmatrix} = \lambda^2 - (a+d)\lambda + (ad - bc) = 0,$$

and the Cayley–Hamilton theorem tells us that A itself satisfies

$$A^2 - (a+d)A + (ad - bc)I = 0$$

where I is the 2×2 identity matrix and the zero on the right side denotes the 2×2 zero matrix. Analogously, for a second rank tensor $\mathbf{A}$ whose characteristic equation is given by (3.16) we have

$$-\mathbf{A}^3 + I_1(\mathbf{A})\mathbf{A}^2 - I_2(\mathbf{A})\mathbf{A} + I_3(\mathbf{A})\mathbf{E} = \mathbf{0}. \tag{3.17}$$

(Here we assume the non-degenerate case.) This nice result permits us to represent $\mathbf{A}^3$ in terms of lower powers of $\mathbf{A}$. Furthermore, we may dot multiply (3.17) by $\mathbf{A}$ and thereby represent $\mathbf{A}^4$ in terms of lower powers of $\mathbf{A}$. It is clear that we could continue in this fashion and eventually express any desired power of $\mathbf{A}$ in terms of $\mathbf{E}$, $\mathbf{A}$, and $\mathbf{A}^2$. This becomes useful in certain applications (e.g., nonlinear elasticity) where functions of tensors are represented approximately by truncated Taylor series.

It is easy to establish the Cayley–Hamilton theorem for the case of a symmetric tensor. Such a tensor $\mathbf{A}$ is represented in diagonal form as[2]

$$\mathbf{A} = \sum_{i=1}^{3} \lambda_i \mathbf{e}_i \mathbf{e}^i.$$

If we dot multiply $\mathbf{A} \cdot \mathbf{A}$ we get

$$\mathbf{A}^2 = \mathbf{A} \cdot \mathbf{A} = \sum_{i=1}^{3} \lambda_i \mathbf{e}_i \mathbf{e}^i \cdot \sum_{j=1}^{3} \lambda_j \mathbf{e}_j \mathbf{e}^j = \sum_{i,j=1}^{3} \lambda_i \mathbf{e}_i \lambda_j \mathbf{e}^j \delta_j^i = \sum_{i=1}^{3} \lambda_i^2 \mathbf{e}_i \mathbf{e}^i.$$

Similarly

$$\mathbf{A}^3 = \mathbf{A} \cdot \mathbf{A} \cdot \mathbf{A} = \sum_{i=1}^{3} \lambda_i^3 \mathbf{e}_i \mathbf{e}^i. \tag{3.18}$$

Let us return to equation (3.16) written for the ith eigenvalue and put it into the expression (3.18):

$$\begin{aligned}
\mathbf{A}^3 &= \sum_{i=1}^{3} [I_1(\mathbf{A})\lambda_i^2 - I_2(\mathbf{A})\lambda_i + I_3(\mathbf{A})] \mathbf{e}_i \mathbf{e}^i \\
&= I_1(\mathbf{A}) \sum_{i=1}^{3} \lambda_i^2 \mathbf{e}_i \mathbf{e}^i - I_2(\mathbf{A}) \sum_{i=1}^{3} \lambda_i \mathbf{e}_i \mathbf{e}^i + I_3(\mathbf{A}) \sum_{i=1}^{3} \mathbf{e}_i \mathbf{e}^i \\
&= I_1(\mathbf{A})\mathbf{A}^2 - I_2(\mathbf{A})\mathbf{A} + I_3(\mathbf{A})\mathbf{E},
\end{aligned}$$

as desired.

Exercise 3.16 Use the Cayley–Hamilton theorem to express $\mathbf{A}^3$ in terms of $\mathbf{A}^2$, $\mathbf{A}$, and $\mathbf{E}$ if $\mathbf{A} = \mathbf{i}_1\mathbf{i}_1 + \mathbf{i}_2\mathbf{i}_1 + \mathbf{i}_2\mathbf{i}_2 + \mathbf{i}_3\mathbf{i}_2$.

Tensors of rotation

A tensor $\mathbf{Q}$ of the second rank is said to be *orthogonal* if it satisfies the equality

$$\mathbf{Q} \cdot \mathbf{Q}^T = \mathbf{Q}^T \cdot \mathbf{Q} = \mathbf{E}.$$

We see that $\mathbf{Q}^T = \mathbf{Q}^{-1}$ for an orthogonal tensor. Furthermore, $\det \mathbf{Q} = \pm 1$. Indeed, $\det \mathbf{Q}$ is determined by the determinant of the matrix of mixed components of $\mathbf{Q}$. Because of the properties of the determinant of a matrix and

[2]The eigenvectors are mutually orthogonal. If we take them as unit vectors, then in this case the dual vectors $\mathbf{e}^i$ coincide with the $\mathbf{e}_i$.

the correspondence between tensors and matrices we have for two tensors $\mathbf{A}$ and $\mathbf{B}$ of the second rank

$$\det(\mathbf{A} \cdot \mathbf{B}) = \det \mathbf{A} \det \mathbf{B}$$

and

$$\det \mathbf{A} = \det \mathbf{A}^T.$$

Thus $1 = \det \mathbf{E} = \det \mathbf{Q} \det \mathbf{Q}^T = (\det \mathbf{Q})^2$ as desired. We call $\mathbf{Q}$ a *proper* orthogonal tensor if $\det \mathbf{Q} = +1$; we call $\mathbf{Q}$ an *improper* orthogonal tensor if $\det \mathbf{Q} = -1$.

Exercise 3.17 (a) Show that the tensor $\mathbf{Q} = -\mathbf{i}_1\mathbf{i}_1 + \mathbf{i}_2\mathbf{i}_2 + \mathbf{i}_3\mathbf{i}_3$ is orthogonal. Is it proper or improper? (b) Show that $q_{ij}q^{kj} = \delta_i^k$ if $\mathbf{Q}$ is orthogonal. (c) Show that if $\mathbf{Q}$ is orthogonal then so is $\mathbf{Q}^n$ for every integer n.

We now consider the orthogonal tensor $\mathbf{Q}$ as an operator in the space of all vectors.

Theorem 3.3 *The operator defined by the orthogonal tensor $\mathbf{Q}$ preserves the magnitudes of vectors and the angles between them.*

Proof. Consider the result of application of $\mathbf{Q}$ to both the multipliers of the inner product $\mathbf{x} \cdot \mathbf{y}$:

$$(\mathbf{Q} \cdot \mathbf{x}) \cdot (\mathbf{Q} \cdot \mathbf{y}) = (\mathbf{x} \cdot \mathbf{Q}^T) \cdot (\mathbf{Q} \cdot \mathbf{y}) = \mathbf{x} \cdot \mathbf{Q}^T \cdot \mathbf{Q} \cdot \mathbf{y} = \mathbf{x} \cdot \mathbf{E} \cdot \mathbf{y} = \mathbf{x} \cdot \mathbf{y}.$$

First we put $\mathbf{y} = \mathbf{x}$ to see that $\mathbf{Q}$ preserves vector magnitudes: $|\mathbf{Q} \cdot \mathbf{x}|^2 = |\mathbf{x}|^2$. Thus, by definition of the dot product in terms of the cosine, angles are also preserved. □

The action of $\mathbf{Q}$ amounts to a rotation of all vectors of the space. (More precisely, this is the case for a proper orthogonal tensor; an improper tensor also causes an axis reflection that changes the "handedness" of the frame.) The situation is analogous to the case of a solid body where the position of a point is defined by a vector beginning at a fixed origin of some frame and ending at the point. Any motion of the solid with a fixed point is a rotation with respect to some axis by some angle. The equations of physics should often be introduced in such a way that they are invariant under rotation of the coordinate frame whose position is not determined in space. Such invariance under rotation should be verified, and in large part this can be done by showing that the application of $\mathbf{Q}$ to all the vectors of the relation does not change the form of the relation.

Let us note that some quantities are always invariant under rotations. One of them is the first invariant $I_1(\mathbf{A})$ of a tensor. This quantity, also called the *trace* of the tensor and denoted $\operatorname{tr}(\mathbf{A})$, is the sum of the diagonal mixed components of $\mathbf{A}$:

$$\operatorname{tr}(\mathbf{A}) = a_{i}^{\cdot i}.$$

The trace can be equivalently determined in the non-coordinate form

$$\operatorname{tr}(\mathbf{A}) = \mathbf{E} \cdot\cdot \mathbf{A}$$

(the reader should check this). To show invariance we consider the tensor

$$\mathbf{Q} \cdot \mathbf{A} \cdot \mathbf{Q}^T = \mathbf{Q} \cdot (a^{ij}\mathbf{e}_i\mathbf{e}_j) \cdot \mathbf{Q}^T = a^{ij}(\mathbf{Q} \cdot \mathbf{e}_i)(\mathbf{e}_j \cdot \mathbf{Q}^T) = a^{ij}(\mathbf{Q} \cdot \mathbf{e}_i)(\mathbf{Q} \cdot \mathbf{e}_j).$$

We see that this is a representation of the "rotated" tensor, derived as the result of applying $\mathbf{Q}$ to each vector of the dyadic components of $\mathbf{A}$. The trace of the rotated tensor is given by

$$\mathbf{E} \cdot\cdot (\mathbf{Q} \cdot \mathbf{A} \cdot \mathbf{Q}^T) = \mathbf{E} \cdot\cdot (\mathbf{Q} \cdot a_{i}^{\cdot j}\mathbf{e}^i\mathbf{e}_j \cdot \mathbf{Q}^T) = \mathbf{E} \cdot\cdot a_{i}^{\cdot j}(\mathbf{Q} \cdot \mathbf{e}^i)(\mathbf{Q} \cdot \mathbf{e}_j).$$

Under the action of $\mathbf{Q}$ the frame $\mathbf{e}_i$ transforms to the frame $\mathbf{Q} \cdot \mathbf{e}_i$, to which the reciprocal basis is $\mathbf{Q} \cdot \mathbf{e}^i$. This means that the unit tensor (the metric tensor!) can be represented, in particular, as

$$\mathbf{E} = (\mathbf{Q} \cdot \mathbf{e}_i)(\mathbf{Q} \cdot \mathbf{e}^i) = (\mathbf{Q} \cdot \mathbf{e}^j)(\mathbf{Q} \cdot \mathbf{e}_j).$$

It follows that

$$\begin{aligned}
\mathbf{E} \cdot\cdot (\mathbf{Q} \cdot \mathbf{A} \cdot \mathbf{Q}^T) &= \left[(\mathbf{Q} \cdot \mathbf{e}^k)(\mathbf{Q} \cdot \mathbf{e}_k)\right] \cdot\cdot a_{i}^{\cdot j}(\mathbf{Q} \cdot \mathbf{e}^i)(\mathbf{Q} \cdot \mathbf{e}_j) \\
&= a_{i}^{\cdot j}\left[(\mathbf{Q} \cdot \mathbf{e}_k) \cdot (\mathbf{Q} \cdot \mathbf{e}^i)\right]\left[(\mathbf{Q} \cdot \mathbf{e}^k) \cdot (\mathbf{Q} \cdot \mathbf{e}_j)\right] \\
&= a_{i}^{\cdot j}\left[(\mathbf{e}_k \cdot \mathbf{Q}^T) \cdot (\mathbf{Q} \cdot \mathbf{e}^i)\right]\left[(\mathbf{e}^k \cdot \mathbf{Q}^T) \cdot (\mathbf{Q} \cdot \mathbf{e}_j)\right] \\
&= a_{i}^{\cdot j}\left[\mathbf{e}_k \cdot \mathbf{Q}^T \cdot \mathbf{Q} \cdot \mathbf{e}^i\right]\left[\mathbf{e}^k \cdot \mathbf{Q}^T \cdot \mathbf{Q} \cdot \mathbf{e}_j\right] \\
&= a_{i}^{\cdot j}\left[\mathbf{e}_k \cdot \mathbf{E} \cdot \mathbf{e}^i\right]\left[\mathbf{e}^k \cdot \mathbf{E} \cdot \mathbf{e}_j\right] \\
&= a_{i}^{\cdot j}\left[\mathbf{e}_k \cdot \mathbf{e}^i\right]\left[\mathbf{e}^k \cdot \mathbf{e}_j\right] \\
&= a_{i}^{\cdot j}\delta_k^i\delta_j^k = a_{k}^{\cdot k} = \operatorname{tr}(\mathbf{A}).
\end{aligned}$$

Another example demonstrates that under the transformation $\mathbf{Q}$ the eigenvalues of a tensor $\mathbf{A}$ remain the same but the eigenvectors $\mathbf{x}_i$ rotate. Indeed, let $\mathbf{A} \cdot \mathbf{x}_i = \lambda_i\mathbf{x}_i$. We demonstrate that $\mathbf{Q} \cdot \mathbf{x}_i$ is an eigenvector of

$\mathbf{Q} \cdot \mathbf{A} \cdot \mathbf{Q}^T$. Thus

$$\begin{aligned}(\mathbf{Q} \cdot \mathbf{A} \cdot \mathbf{Q}^T) \cdot (\mathbf{Q} \cdot \mathbf{x}_i) &= \mathbf{Q} \cdot \mathbf{A} \cdot \mathbf{Q}^T \cdot \mathbf{Q} \cdot \mathbf{x}_i \\ &= \mathbf{Q} \cdot \mathbf{A} \cdot \mathbf{E} \cdot \mathbf{x}_i \\ &= \mathbf{Q} \cdot (\mathbf{A} \cdot \mathbf{x}_i) \\ &= \mathbf{Q} \cdot (\lambda_i \mathbf{x}_i) \\ &= \lambda_i (\mathbf{Q} \cdot \mathbf{x}_i).\end{aligned}$$

Let $\mathbf{e}$ be an axis of rotation defined by an orthogonal tensor and ω the angle of rotation. It can be shown that the orthogonal tensor has the representation

$$\mathbf{Q} = \mathbf{E} \cos\omega + (1 - \cos\omega)\mathbf{e}\mathbf{e} - \mathbf{e} \times \mathbf{E} \sin\omega.$$

Polar decomposition

A second rank tensor $\mathbf{A}$ is nonsingular if

$$\det A \neq 0$$

where A is the matrix of mixed components of $\mathbf{A}$. It is possible to express such a tensor as a product of a symmetric tensor and another tensor. A statement of this result, known as the *polar decomposition theorem*, requires that we introduce some additional terminology.

A symmetric tensor is said to be *positive definite* if its eigenvalues are all positive. By orthogonal decomposition such a tensor $\mathbf{S}$ may be written in the form

$$\mathbf{S} = \lambda_1 \mathbf{e}_1 \mathbf{e}_1 + \lambda_2 \mathbf{e}_2 \mathbf{e}_2 + \lambda_3 \mathbf{e}_3 \mathbf{e}_3$$

where each $\lambda_i > 0$. If $\mathbf{A}$ is nonsingular then the tensor $\mathbf{A} \cdot \mathbf{A}^T$ is symmetric and positive definite. Symmetry follows from the equation

$$(\mathbf{A} \cdot \mathbf{A}^T)^T = (\mathbf{A}^T)^T \cdot \mathbf{A}^T = \mathbf{A} \cdot \mathbf{A}^T.$$

To see positive definiteness, we begin with the definition

$$(\mathbf{A} \cdot \mathbf{A}^T) \cdot \mathbf{x} = \lambda \mathbf{x}$$

of an eigenvalue λ and dot with $\mathbf{x}$ from the left to get

$$\lambda = \frac{\mathbf{x} \cdot (\mathbf{A} \cdot \mathbf{A}^T) \cdot \mathbf{x}}{|\mathbf{x}|^2}.$$

The numerator is positive because

$$\mathbf{x} \cdot (\mathbf{A} \cdot \mathbf{A}^T) \cdot \mathbf{x} = (\mathbf{x} \cdot \mathbf{A}) \cdot (\mathbf{A}^T \cdot \mathbf{x}) = (\mathbf{x} \cdot \mathbf{A}) \cdot (\mathbf{x} \cdot \mathbf{A}) = |\mathbf{x} \cdot \mathbf{A}|^2.$$

With these facts in hand we may turn to our main result:

Theorem 3.4 *Any nonsingular tensor* $\mathbf{A}$ *of rank two may be written as a product of an orthogonal tensor and a positive definite symmetric tensor. The decomposition may be done in two ways: as a* left polar decomposition

$$\mathbf{A} = \mathbf{S} \cdot \mathbf{Q} \tag{3.19}$$

or as a right polar decomposition

$$\mathbf{A} = \mathbf{Q} \cdot \mathbf{S}'. \tag{3.20}$$

Here $\mathbf{Q}$ *is an orthogonal tensor of rank two, and* $\mathbf{S}$ *and* $\mathbf{S}'$ *are positive definite and symmetric.*

Proof. Because $\mathbf{A} \cdot \mathbf{A}^T$ is positive definite and symmetric we have

$$\mathbf{A} \cdot \mathbf{A}^T = \lambda_1 \mathbf{e}_1\mathbf{e}_1 + \lambda_2 \mathbf{e}_2\mathbf{e}_2 + \lambda_3 \mathbf{e}_3\mathbf{e}_3$$

and can define

$$\mathbf{S} \equiv (\mathbf{A} \cdot \mathbf{A}^T)^{1/2} = \sqrt{\lambda_1}\, \mathbf{e}_1\mathbf{e}_1 + \sqrt{\lambda_2}\, \mathbf{e}_2\mathbf{e}_2 + \sqrt{\lambda_3}\, \mathbf{e}_3\mathbf{e}_3$$

since the λ_i are positive. We see that $\mathbf{S}^{-1}$ exists because

$$\mathbf{S}^{-1} = \frac{1}{\sqrt{\lambda_1}}\mathbf{e}_1\mathbf{e}_1 + \frac{1}{\sqrt{\lambda_2}}\mathbf{e}_2\mathbf{e}_2 + \frac{1}{\sqrt{\lambda_3}}\mathbf{e}_3\mathbf{e}_3$$

where the λ_i are positive. Now we set

$$\mathbf{Q} \equiv \mathbf{S}^{-1} \cdot \mathbf{A}.$$

To see that $\mathbf{Q}$ is orthogonal we write

$$\begin{aligned}
\mathbf{Q} \cdot \mathbf{Q}^T &= (\mathbf{S}^{-1} \cdot \mathbf{A}) \cdot (\mathbf{S}^{-1} \cdot \mathbf{A})^T \\
&= (\mathbf{S}^{-1} \cdot \mathbf{A}) \cdot (\mathbf{A}^T \cdot (\mathbf{S}^{-1})^T) \\
&= \mathbf{S}^{-1} \cdot (\mathbf{A} \cdot \mathbf{A}^T) \cdot (\mathbf{S}^T)^{-1} \\
&= \mathbf{S}^{-1} \cdot \mathbf{S}^2 \cdot \mathbf{S}^{-1} \\
&= \mathbf{E}.
\end{aligned}$$

Thus we have expressed $\mathbf{A} = \mathbf{S} \cdot \mathbf{Q}$ as in (3.19). The validity of (3.20) follows from defining $\mathbf{S}' \equiv \mathbf{Q}^T \cdot \mathbf{S} \cdot \mathbf{Q}$. □

Exercise 3.18 We have called a tensor positive definite if its eigenvalues are all positive. An alternative definition is that $\mathbf{A}$ is positive definite if $\mathbf{x} \cdot (\mathbf{A} \cdot \mathbf{x}) > 0$ for all $\mathbf{x} \neq 0$. Explain.

As an application of polar decomposition, let us show that a nonsingular tensor $\mathbf{A}$ operates on the position vectors touching a sphere to produce position vectors touching an ellipsoid. If $\mathbf{r}$ locates any point on the unit sphere then

$$\mathbf{r} \cdot \mathbf{r} = 1. \tag{3.21}$$

Let $\mathbf{x}$ be the image of $\mathbf{r}$ under $\mathbf{A}$:

$$\mathbf{x} = \mathbf{A} \cdot \mathbf{r}.$$

Then $\mathbf{r} = \mathbf{A}^{-1} \cdot \mathbf{x}$, and substitution into (3.21) along with (3.19) gives

$$\begin{aligned} 1 &= [(\mathbf{S} \cdot \mathbf{Q})^{-1} \cdot \mathbf{x}] \cdot [(\mathbf{S} \cdot \mathbf{Q})^{-1} \cdot \mathbf{x}] \\ &= \mathbf{x} \cdot \{[(\mathbf{S} \cdot \mathbf{Q})^{-1}]^T \cdot (\mathbf{S} \cdot \mathbf{Q})^{-1}\} \cdot \mathbf{x} \\ &= \mathbf{x} \cdot \{\mathbf{S}^{-1} \cdot \mathbf{S}^{-1}\} \cdot \mathbf{x} \end{aligned}$$

(the reader can supply the missing details). Expansion in a Cartesian frame using

$$\mathbf{x} = \sum_{i=1}^{3} x_i \mathbf{i}_i, \qquad \mathbf{S}^{-1} = \sum_{j=1}^{3} \frac{1}{\sqrt{\lambda_j}} \mathbf{i}_j \mathbf{i}_j,$$

reduces this equation to the form

$$\sum_{i=1}^{3} \frac{1}{\lambda_i} x_i^2 = 1,$$

which is the equation of an ellipsoid.

Polar decomposition provides the background for the introduction of measures of deformation in nonlinear continuum mechanics.

Isotropic tensors and isotropic scalar functions

A tensor is said to be *isotropic* if its individual components are invariant under rotations of the frame. It can be shown that the only isotropic tensor of the second rank is $\lambda \mathbf{E}$.

A scalar function of a tensor is called isotropic if it keeps its form under any orthogonal transformation of the space (basis). If it keeps the form under some subgroup of orthogonal transformations we get functions that are invariant under them. The reader will find applications of this idea in books on crystallography and elasticity.

Deviator and ball tensor representation

For $\mathbf{A}$ we can introduce the representation

$$\mathbf{A} = \frac{1}{3} I_1(\mathbf{A})\mathbf{E} + \operatorname{dev} \mathbf{Q}.$$

Such a representation is found useful in the theory of elasticity. The tensor $\operatorname{dev} \mathbf{Q}$ is defined by the above equality. It has the same eigenvectors as $\mathbf{A}$, but eigenvalues that differ from the eigenvalues of $\mathbf{A}$ by $1/3 I_1(\mathbf{A})$:

$$\tilde{\lambda}_i = \lambda_i - \frac{1}{3} \operatorname{tr} \mathbf{A}.$$

3.6 Extending the Dyad Idea

The introduction of tensors of the third rank can be done in a way that parallels the introduction of dyads in § 3.1. Utilizing the tensor product as before, we introduce *triad* quantities of the type

$$\mathbf{R} = \mathbf{abc}$$

where $\mathbf{a}$, $\mathbf{b}$, and $\mathbf{c}$ are vectors. Expanding these vectors in terms of a basis $\mathbf{e}_i$ we obtain

$$\mathbf{R} = a^i b^j c^k \mathbf{e}_i \mathbf{e}_j \mathbf{e}_k.$$

We then consider a linear space whose basis is the set of 27 quantities $\mathbf{e}_i \mathbf{e}_j \mathbf{e}_k$, and call it the space of tensors of the third rank. We continue to refer to the numerical values $a^i b^j c^k$ as the tensor's components. A general element of this space, a tensor $\mathbf{R}$ of rank three, has the representation

$$\mathbf{R} = r^{ijk} \mathbf{e}_i \mathbf{e}_j \mathbf{e}_k. \tag{3.22}$$

The property of objectivity of $\mathbf{R}$ remains paramount, leading to the requirement that the components transform appropriately when we change the frame. The now familiar procedure of setting

$$\tilde{r}^{ijk} \tilde{\mathbf{e}}_i \tilde{\mathbf{e}}_j \tilde{\mathbf{e}}_k = r^{mnp} \mathbf{e}_m \mathbf{e}_n \mathbf{e}_p$$

under the change of frame

$$\mathbf{e}_i = A_i^j \tilde{\mathbf{e}}_j$$

gives

$$\tilde{r}^{ijk} = r^{mnp} A_m^i A_n^j A_p^k$$

— a direct extension of (3.8). As an alternative to the representation (3.22) in contravariant components, we could use the covariant-type representation

$$\mathbf{R} = r_{ijk} \mathbf{e}^i \mathbf{e}^j \mathbf{e}^k$$

or either of the mixed representations

$$\mathbf{R} = r^{ij}_{\cdot\cdot k} \mathbf{e}_i \mathbf{e}_j \mathbf{e}^k, \qquad \mathbf{R} = r^{i}_{\cdot jk} \mathbf{e}_i \mathbf{e}^j \mathbf{e}^k.$$

These necessitate the respective transformation laws

$$\tilde{r}_{ijk} = r_{mnp} \tilde{A}_i^m \tilde{A}_j^n \tilde{A}_k^p$$

and

$$\tilde{r}^{ij}_{\cdot\cdot k} = r^{mn}_{\cdot\cdot p} A_m^i A_n^j \tilde{A}_k^p, \qquad \tilde{r}^{i}_{\cdot jk} = r^{m}_{\cdot np} A_m^i \tilde{A}_j^n \tilde{A}_k^p,$$

as is easily verified.

Dot products involving triads follow familiar rules. The dot product of a triad with a vector is given by the formula

$$\mathbf{abc} \cdot \mathbf{x} = \mathbf{ab}(\mathbf{c} \cdot \mathbf{x}),$$

while the double dot product of a triad with a dyad is given by

$$\mathbf{abc} \cdot\cdot\, \mathbf{xy} = \mathbf{a}(\mathbf{c} \cdot \mathbf{x})(\mathbf{b} \cdot \mathbf{y}).$$

One may also define a triple dot product of a triad with another triad:

$$\mathbf{abc} \cdots \mathbf{xyz} = (\mathbf{c} \cdot \mathbf{x})(\mathbf{b} \cdot \mathbf{y})(\mathbf{a} \cdot \mathbf{z}).$$

The reader has surmised by now that the rank of a tensor is always equal to the number of free indices needed to specify its components. A vector, for instance, is a tensor of rank one. In Chapter 2 we also met a quantity whose components are specified by three indices: ϵ_{ijk}. This quantity, which arose naturally in our discussion of the vector cross product, is known as the *Levi–Civita tensor* and is given by

$$\boldsymbol{\mathcal{E}} = \epsilon_{ijk}\mathbf{e}^i\mathbf{e}^j\mathbf{e}^k.$$

Exercise 3.19 Verify the following formulas for operations involving the Levi–Civita tensor:

(a) $\boldsymbol{\mathcal{E}} \cdots \mathbf{zyx} = \mathbf{x} \cdot (\mathbf{y} \times \mathbf{z})$,
(b) $\boldsymbol{\mathcal{E}} \cdot\cdot\, \mathbf{xy} = \mathbf{y} \times \mathbf{x}$,
(c) $\boldsymbol{\mathcal{E}} \cdot \mathbf{x} = -\mathbf{x}\times$.

Note: The notation of (c) may require some explanation. The result of applying the third rank tensor $\boldsymbol{\mathcal{E}}$ to a vector $\mathbf{x}$ is a second rank tensor $\boldsymbol{\mathcal{E}} \cdot \mathbf{x}$. When $\boldsymbol{\mathcal{E}} \cdot \mathbf{x}$ is applied to another vector $\mathbf{y}$, it becomes equivalent to the cross product $\mathbf{y} \times \mathbf{x}$. This is the meaning of the right side of (c). Although the notation is awkward, it is rare that the action of some tensor can be described as the action of two vectors, and the development of a special notation is unwarranted.

3.7 Tensors of the Fourth and Higher Ranks

We can obviously extend the present treatment to tensors of any desired rank. Let us illustrate the essential points using tensors of rank four.

A tensor $\mathbf{C}$ of rank four can be represented by several types of components:

$$c^{ijkl}, \qquad c_i^{\cdot jkl}, \qquad c_{ij}^{\cdot\cdot kl}, \qquad c_{ijk}^{\cdot\cdot\cdot l}, \qquad c_{ijkl}.$$

The first and last are purely contravariant and purely covariant, respectively, while the other three are mixed with indices in various positions. As before these components represent $\mathbf{C}$ with respect to various bases; for instance,

$$\mathbf{C} = c^{ijkl}\mathbf{e}_i\mathbf{e}_j\mathbf{e}_k\mathbf{e}_l.$$

Dot products with vectors can be taken as before: the rule is that we simply dot multiply the basis vectors positioned nearest to the dot. Carrying out such operations we may obtain results of various kinds. A dot product of a fourth rank tensor with a vector gives, for example,

$$\mathbf{R} = \mathbf{C} \cdot \mathbf{x} = c^{ijkl}\mathbf{e}_i\mathbf{e}_j\mathbf{e}_k\mathbf{e}_l \cdot x_m\mathbf{e}^m = c^{ijkl}\mathbf{e}_i\mathbf{e}_j\mathbf{e}_k\delta_l^m x_m = c^{ijkl}x_l\mathbf{e}_i\mathbf{e}_j\mathbf{e}_k$$

— a tensor of rank three. Similarly, a dot product between a third rank tensor and a vector gives a second rank tensor. Wherever the dot product is utilized, it continues to enjoy the linearity properties that were stated earlier.

Double dot products also appear in applications. For example, in elasticity one encounters double dot products between the tensor of elastic constants and the strain tensor. A double dot product between the stress and strain tensors gives the double density of internal energy of a linearly elastic body. In generalized form Hooke's law becomes

$$\boldsymbol{\sigma} = \mathbf{C} \cdot\cdot\, \boldsymbol{\epsilon}$$

where $\boldsymbol{\sigma}$ is the stress tensor, $\mathbf{C}$ is the tensor of elastic constants, and $\boldsymbol{\epsilon}$ is the strain tensor. The density of the function of internal (elastic) energy in linear elasticity is

$$\frac{1}{2}\boldsymbol{\sigma} \cdot\cdot\, \boldsymbol{\epsilon} = \frac{1}{2}(\mathbf{C} \cdot\cdot\, \boldsymbol{\epsilon}) \cdot\cdot\, \boldsymbol{\epsilon}.$$

This last expression can be put in a more symmetrical form

$$\frac{1}{2}\boldsymbol{\epsilon} \cdot\cdot\, \mathbf{C} \cdot\cdot\, \boldsymbol{\epsilon}$$

in which the result does not depend on the order of operations. The form of the elastic energy should be positive, which means that it is greater than zero for any nonzero $\boldsymbol{\epsilon}$. The components of $\mathbf{C}$ have a certain symmetry, so the number of independent parameters is no more than 21. For an isotropic material these depend only on two elastic constants E and ν (the Poisson ratio). There are other types of materials (orthotropic, etc.) for which the number of independent constants is greater than this.

Chapter 4

Tensor Fields

4.1 Vector Fields

The state of a point (more precisely of an infinitely small volume) within a natural object is frequently characterized by a vector or a tensor. Hence inside a spatial object there arises what we call a *vector* or *tensor field*. As a rule these fields are governed by simultaneous partial differential equations. Such equations are usually derived using a Cartesian space frame. In this frame, the operations of calculus closely parallel those of one dimensional analysis: the differentiation of a vector function proceeds on a component by component basis, for example. However, it is often convenient to introduce so-called curvilinear coordinates in the body, in terms of which the problem formulation is simpler. In this way we get a frame that changes from point to point, and component-wise differentiation is not enough to characterize the change of the vector function. Thus we need to develop the apparatus of calculus for vector and tensor functions when the frame in the object is changeable. Another reason for introducing these tools is the objectivity of the laws of nature: we must be able to formulate frame independent statements of these laws. Finally, there is an aesthetic reason: in non-coordinate form many statements of mathematical physics look much less cumbersome than their counterparts stated in terms of coordinates. It is said that beauty governs the world; although this is not absolutely true, most students would prefer a short and beautiful statement to a nightmarish formula taking half a page.

The position of a point in space is characterized by three numbers called coordinates of the point. Coordinate systems in common use include the Cartesian, cylindrical, and spherical systems. The first of these differs from the latter two in an important respect: the frame vectors of a Cartesian system are unique, while the frame vectors for the curvilinear systems change from point to point. In a general coordinate system the position of a point in space is determined uniquely by three numbers q^1, q^2, q^3. These are referred to as *curvilinear coordinates* if the frame is not Cartesian. See Fig. 4.1.

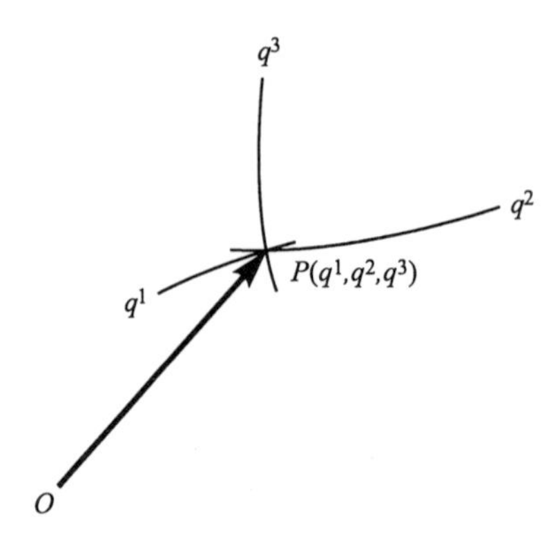

Fig. 4.1 Curvilinear coordinates.

If we fix two of the coordinates and change the third, we get a line in space called a *coordinate line.* The Cartesian coordinates x^1, x^2, x^3 of a point can be determined through the general curvilinear coordinates by relations of the general form

$$x^i = x^i(q^1, q^2, q^3), \quad i = 1, 2, 3. \tag{4.1}$$

Except for some set of singular points in space, the correspondence (4.1) is one to one. We suppose the functions $x^i = x^i(q^1, q^2, q^3)$ to be smooth (continuously differentiable). In this case the local one-to-one correspondence is provided by the requirement that the Jacobian

$$\left| \frac{\partial x^i}{\partial q^j} \right| \neq 0. \tag{4.2}$$

Fixing some origin O, we characterize the position of a point $P(q^1, q^2, q^3)$ by a vector $\mathbf{r}$ connecting the points O and P:

$$\mathbf{r} = \mathbf{r}(q^1, q^2, q^3).$$

When a point moves along a coordinate line, along the q^1 line for instance, the end of its position vector moves along this line and the difference

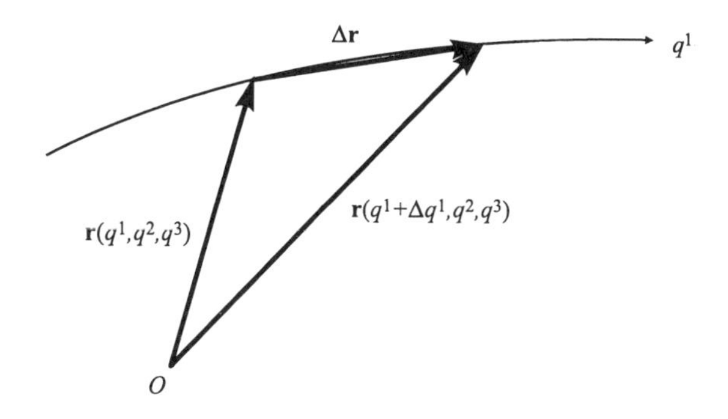

Fig. 4.2 Generation of a local tangent vector to a coordinate line.

vector

$$\Delta \mathbf{r} = \mathbf{r}(q^1 + \Delta q^1, q^2, q^3) - \mathbf{r}(q^1, q^2, q^3)$$

is directed along the chord (Fig. 4.2). The smaller the value of Δq^1, the closer $\Delta \mathbf{r}$ is to being tangent to the q^1 line. The limit as $\Delta q^1 \to 0$ of the ratio $\Delta \mathbf{r}/\Delta q^1$ is a vector

$$\frac{\partial \mathbf{r}(q^1, q^2, q^3)}{\partial q^1} = \mathbf{r}_1$$

tangent to the q^1 coordinate line. At the same point (q^1, q^2, q^3) we can introduce two other vectors

$$\mathbf{r}_2 = \frac{\partial \mathbf{r}(q^1, q^2, q^3)}{\partial q^2}, \qquad \mathbf{r}_3 = \frac{\partial \mathbf{r}(q^1, q^2, q^3)}{\partial q^3},$$

tangent to the q^2 and q^3 coordinate lines, respectively. If the coordinate point (q^1, q^2, q^3) is not singular then the vectors $\mathbf{r}_i$ are non-coplanar and therefore constitute a frame triad. The mixed product $\mathbf{r}_1 \cdot (\mathbf{r}_2 \times \mathbf{r}_3)$ is the volume of the frame parallelepiped; renumbering the coordinates if necessary, we obtain from this the same expression for the Jacobian (4.2). Let us denote it

$$\sqrt{g} = \mathbf{r}_1 \cdot (\mathbf{r}_2 \times \mathbf{r}_3) = \left| \frac{\partial x^i}{\partial q^j} \right| .$$

Unlike the previous chapters in which all frame vectors were constant, we now deal with frame vectors that change from point to point. However, at each point we can repeat our prior reasoning. In particular, let us introduce the reciprocal basis $\mathbf{r}^i$ by the relation

$$\mathbf{r}^i \cdot \mathbf{r}_j = \delta^i_j.$$

By the previous chapter we define the metric coefficients

$$g_{ij} = \mathbf{r}_i \cdot \mathbf{r}_j, \qquad g^{ij} = \mathbf{r}^i \cdot \mathbf{r}^j, \qquad g^j_i = \mathbf{r}_i \cdot \mathbf{r}^j = \delta^j_i,$$

which are the components of the metric tensor $\mathbf{E}$ at each point in space.

Above we differentiated a vector function $\mathbf{r}(q^1, q^2, q^3)$. Note that the rules for differentiating vector functions are quite similar to those for differentiating usual functions. For brevity we consider the case of fields depending on one variable t. Let $\mathbf{e}_1(t)$ and $\mathbf{e}_2(t)$ be continuously differentiable at some finite t, which means that for $i = 1, 2$ there exist

$$\frac{d\mathbf{e}_i(t)}{dt} = \lim_{\Delta t \to 0} \frac{\mathbf{e}_i(t + \Delta t) - \mathbf{e}_i(t)}{\Delta t}.$$

It is easily seen that

$$\frac{d(\mathbf{e}_1(t) + \mathbf{e}_1(t))}{dt} = \frac{d\mathbf{e}_1(t)}{dt} + \frac{d\mathbf{e}_2(t)}{dt}.$$

Indeed,

$$\begin{aligned} \frac{d(\mathbf{e}_1(t) + \mathbf{e}_1(t))}{dt} &= \lim_{\Delta t \to 0} \frac{\mathbf{e}_1(t + \Delta t) - \mathbf{e}_1(t) + \mathbf{e}_2(t + \Delta t) - \mathbf{e}_2(t)}{\Delta t} \\ &= \lim_{\Delta t \to 0} \frac{\mathbf{e}_1(t + \Delta t) - \mathbf{e}_1(t)}{\Delta t} + \lim_{\Delta t \to 0} \frac{\mathbf{e}_2(t + \Delta t) - \mathbf{e}_2(t)}{\Delta t} \\ &= \frac{d\mathbf{e}_1(t)}{dt} + \frac{d\mathbf{e}_2(t)}{dt}. \end{aligned}$$

Similarly, for a constant c

$$\frac{d(c\mathbf{e}_1(t))}{dt} = c\frac{d\mathbf{e}_1(t)}{dt}$$

which is proved like the previous property. Finally, the product of a scalar function $f(t)$ and a vector function $\mathbf{e}(t)$ is differentiated by a rule similar to the formula for differentiating a product of scalar functions:

$$\frac{d(f(t)\mathbf{e}(t))}{dt} = \frac{df(t)}{dt}\mathbf{e}(t) + f(t)\frac{d\mathbf{e}(t)}{dt}.$$

The reader can adapt the proof for the scalar case almost word for word. The rules for partial differentiation of vector functions in several variables look quite similar and we leave their formulation to the reader. Representing vectors in Cartesian coordinates it is easy to see the validity of the following formulas:

$$\frac{d}{dt}(\mathbf{e}_1(t)\cdot\mathbf{e}_2(t)) = \mathbf{e}_1'(t)\cdot\mathbf{e}_2(t) + \mathbf{e}_1(t)\cdot\mathbf{e}_2'(t),$$

$$\frac{d}{dt}(\mathbf{e}_1(t)\times\mathbf{e}_2(t)) = \mathbf{e}_1'(t)\times\mathbf{e}_2(t) + \mathbf{e}_1(t)\times\mathbf{e}_2'(t).$$

Exercise 4.1 (a) Differentiate $\mathbf{e}_1(t)\cdot\mathbf{e}_2(t)$ and $\mathbf{e}_1(t)\times\mathbf{e}_2(t)$ with respect to t if

$$\mathbf{e}_1(t) = \mathbf{i}_1 e^{-t} + \mathbf{i}_2, \qquad \mathbf{e}_2(t) = -\mathbf{i}_1\sin^2 t + \mathbf{i}_2 e^{-t}.$$

(b) Show that $[\mathbf{e}(t)\times\mathbf{e}'(t)]' = \mathbf{e}(t)\times\mathbf{e}''(t)$ for any differentiable vector function $\mathbf{e}(t)$.

Exercise 4.2 For the mixed product

$$[\mathbf{e}_1(t)\,\mathbf{e}_2(t)\,\mathbf{e}_3(t)] = (\mathbf{e}_1(t)\times\mathbf{e}_2(t))\cdot\mathbf{e}_3(t)$$

show that

$$\begin{aligned}\frac{d}{dt}[\mathbf{e}_1(t)\,\mathbf{e}_2(t)\,\mathbf{e}_3(t)] = {} & [\mathbf{e}_1'(t)\,\mathbf{e}_2(t)\,\mathbf{e}_3(t)] + \\ & + [\mathbf{e}_1(t)\,\mathbf{e}_2'(t)\,\mathbf{e}_3(t)] + \\ & + [\mathbf{e}_1(t)\,\mathbf{e}_2(t)\,\mathbf{e}_3'(t)].\end{aligned}$$

Cylindrical coordinates

In the cylindrical coordinate system

$$(q^1, q^2, q^3) = (\rho, \phi, z)$$

where ρ is the radial distance from the z-axis and ϕ is the so-called azimuthal angle (Fig. 4.3). Using the expression

$$\mathbf{r} = \hat{\mathbf{x}}x + \hat{\mathbf{y}}y + \hat{\mathbf{z}}z$$

for the position vector in rectangular coordinates and the coordinate transformation formulas

$$x = \rho\cos\phi, \qquad y = \rho\sin\phi, \qquad z = z,$$

we have

$$\mathbf{r}(\rho,\phi,z) = \hat{\mathbf{x}}\rho\cos\phi + \hat{\mathbf{y}}\rho\sin\phi + \hat{\mathbf{z}}z.$$

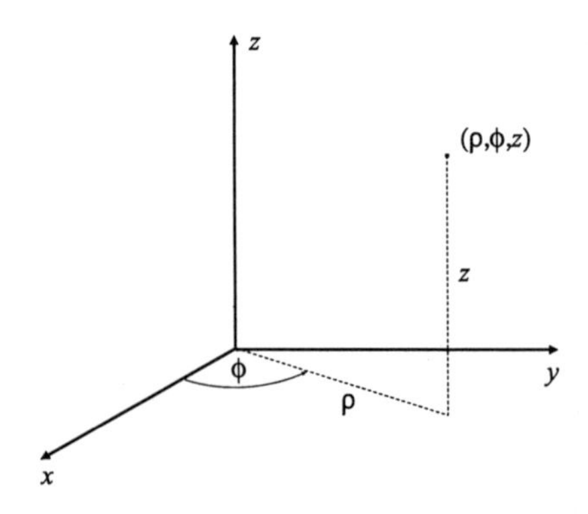

Fig. 4.3 Cylindrical coordinate system.

Then

$$\mathbf{r}_1 = \frac{\partial \mathbf{r}(\rho, \phi, z)}{\partial \rho} = \hat{\mathbf{x}} \cos\phi + \hat{\mathbf{y}} \sin\phi.$$

Similarly we compute

$$\mathbf{r}_2 = \frac{\partial \mathbf{r}(\rho, \phi, z)}{\partial \phi} = -\hat{\mathbf{x}}\rho \sin\phi + \hat{\mathbf{y}}\rho \cos\phi, \qquad \mathbf{r}_3 = \frac{\partial \mathbf{r}(\rho, \phi, z)}{\partial z} = \hat{\mathbf{z}}.$$

Then

$$\begin{aligned}
\sqrt{g} &= \mathbf{r}_1 \cdot (\mathbf{r}_2 \times \mathbf{r}_3) \\
&= (\hat{\mathbf{x}} \cos\phi + \hat{\mathbf{y}} \sin\phi) \cdot [(-\hat{\mathbf{x}}\rho \sin\phi + \hat{\mathbf{y}}\rho \cos\phi) \times \hat{\mathbf{z}}] \\
&= (\hat{\mathbf{x}} \cos\phi + \hat{\mathbf{y}} \sin\phi) \cdot (\hat{\mathbf{y}}\rho \sin\phi + \hat{\mathbf{x}}\rho \cos\phi) \\
&= \rho(\cos^2\phi + \sin^2\phi) \\
&= \rho.
\end{aligned}$$

It is easy to verify that the frame vectors $\mathbf{r}_1, \mathbf{r}_2, \mathbf{r}_3$ are mutually perpendicular. For instance, we have

$$\begin{aligned}
\mathbf{r}_1 \cdot \mathbf{r}_2 &= (\hat{\mathbf{x}} \cos\phi + \hat{\mathbf{y}} \sin\phi) \cdot (-\hat{\mathbf{x}}\rho \sin\phi + \hat{\mathbf{y}}\rho \cos\phi) \\
&= -\rho \cos\phi \sin\phi + \rho \cos\phi \sin\phi = 0.
\end{aligned}$$

By direct computation we also find that

$$|\mathbf{r}_1| = 1, \qquad |\mathbf{r}_2| = \rho, \qquad |\mathbf{r}_3| = 1.$$

These facts may be used to construct the reciprocal basis as

$$\mathbf{r}^1 = \mathbf{r}_1, \qquad \mathbf{r}^2 = \mathbf{r}_2/\rho^2, \qquad \mathbf{r}^3 = \mathbf{r}_3.$$

The various metric coefficients for the cylindrical frame are easily computed, and are

$$(g_{ij}) = \begin{pmatrix} 1 & 0 & 0 \\ 0 & \rho^2 & 0 \\ 0 & 0 & 1 \end{pmatrix} \qquad (g^{ij}) = \begin{pmatrix} 1 & 0 & 0 \\ 0 & 1/\rho^2 & 0 \\ 0 & 0 & 1 \end{pmatrix}$$

Commonly the *unit basis* vectors $\hat{\boldsymbol{\rho}}$, $\hat{\boldsymbol{\phi}}$, and $\hat{\mathbf{z}}$ are introduced in cylindrical coordinates. These are unit vectors along the directions of $\mathbf{r}_1$, $\mathbf{r}_2$, and $\mathbf{r}_3$, respectively, at the point of interest. Given any vector $\mathbf{v}$ at a point (ρ, ϕ, z) in space, it is conventional in applications to write

$$\mathbf{v} = \hat{\boldsymbol{\rho}} v_\rho + \hat{\boldsymbol{\phi}} v_\phi + \hat{\mathbf{z}} v_z.$$

The quantities v_ρ, v_ϕ, and v_z are known as the *physical components* of $\mathbf{v}$. (For a vector, the physical components are the projections of the vector on the directions of the corresponding frame vectors.[1]) But since we can also express $\mathbf{v}$ in the forms

$$\mathbf{v} = v^1\mathbf{r}_1 + v^2\mathbf{r}_2 + v^3\mathbf{r}_3 = v_1\mathbf{r}^1 + v_2\mathbf{r}^2 + v_3\mathbf{r}^3,$$

we can identify the contravariant and covariant components of $\mathbf{v}$ from our previous expressions. They are

$$(v^1, v^2, v^3) = (v_\rho, v_\phi/\rho, v_z), \qquad (v_1, v_2, v_3) = (v_\rho, v_\phi\rho, v_z).$$

We see that expression of $\mathbf{v}$ in terms of the basis $(\hat{\boldsymbol{\rho}}, \hat{\boldsymbol{\phi}}, \hat{\mathbf{z}})$ yields components (v_ρ, v_ϕ, v_z) that are neither contravariant nor covariant.

[1]Mathematicians often prefer to deal with dimensionless quantities. But in applications many quantities have physical dimensions (e.g., N/m^2 for a stress). When we introduce coordinates in space some can also have dimensions; in spherical coordinates, for instance, r has the dimension of length whereas θ and ϕ are dimensionless. Since $\mathbf{r}$ has the length dimension, the frame vectors corresponding to θ and ϕ must have the length dimension, whereas the frame vector for r is dimensionless. In this way some components of the metric tensor may have dimensions and, when involved in transformation formulas, may thus introduce not only numerical changes but also dimensional changes in the terms so that it becomes possible to add such quantities. When one uses physical components of vectors and tensors they take the dimension of the corresponding tensor in full, and the frame vectors become dimensionless.

Exercise 4.3 Show that the acceleration of a particle in plane polar coordinates (ρ, ϕ) is given by

$$\frac{d^2\mathbf{r}}{dt^2} = \hat{\rho}\left[\frac{d^2\rho}{dt^2} - \rho\left(\frac{d\phi}{dt}\right)^2\right] + \hat{\phi}\left(\rho\frac{d^2\phi}{dt^2} + 2\frac{d\rho}{dt}\frac{d\phi}{dt}\right).$$

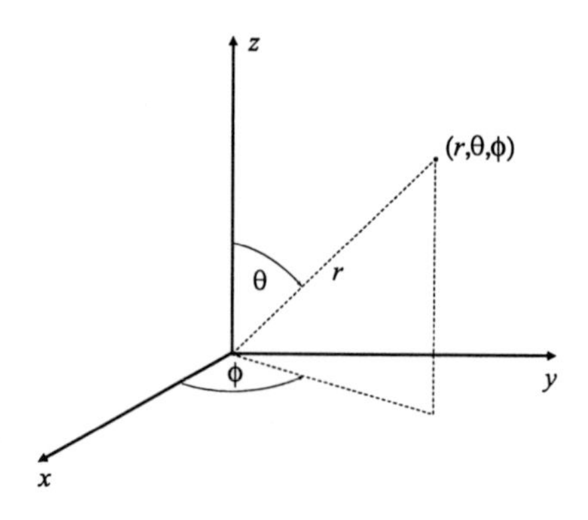

Fig. 4.4 Spherical coordinate system.

Spherical coordinates

In the spherical coordinate system we have

$$(q^1, q^2, q^3) = (r, \theta, \phi),$$

where r is the radial distance from the origin, ϕ is the azimuthal angle, and θ is the so-called polar angle (Fig. 4.4). The coordinate transformation formulas

$$x = r\sin\theta\cos\phi, \qquad y = r\sin\theta\sin\phi, \qquad z = r\cos\theta,$$

give

$$\mathbf{r}(r, \theta, \phi) = \hat{\mathbf{x}}r\sin\theta\cos\phi + \hat{\mathbf{y}}r\sin\theta\sin\phi + \hat{\mathbf{z}}r\cos\theta.$$

Then

$$\mathbf{r}_1 = \frac{\partial \mathbf{r}(r,\theta,\phi)}{\partial r} = \hat{\mathbf{x}} \sin\theta\cos\phi + \hat{\mathbf{y}}\sin\theta\sin\phi + \hat{\mathbf{z}}\cos\theta,$$
$$\mathbf{r}_2 = \frac{\partial \mathbf{r}(r,\theta,\phi)}{\partial \theta} = \hat{\mathbf{x}} r\cos\theta\cos\phi + \hat{\mathbf{y}} r\cos\theta\sin\phi - \hat{\mathbf{z}} r\sin\theta,$$
$$\mathbf{r}_3 = \frac{\partial \mathbf{r}(r,\theta,\phi)}{\partial \phi} = -\hat{\mathbf{x}} r\sin\theta\sin\phi + \hat{\mathbf{y}} r\sin\theta\cos\phi.$$

In this case

$$\sqrt{g} = \begin{vmatrix} \sin\theta\cos\phi & \sin\theta\sin\phi & \cos\theta \\ r\cos\theta\cos\phi & r\cos\theta\sin\phi & -r\sin\theta \\ -r\sin\theta\sin\phi & r\sin\theta\cos\phi & 0 \end{vmatrix} = r^2\sin\theta.$$

In this system we find that

$$|\mathbf{r}_1| = 1, \qquad |\mathbf{r}_2| = r, \qquad |\mathbf{r}_3| = r\sin\theta,$$

and the frame vectors are again mutually orthogonal. Hence we have

$$\mathbf{r}^1 = \mathbf{r}_1, \qquad \mathbf{r}^2 = \mathbf{r}_2/r^2, \qquad \mathbf{r}^3 = \mathbf{r}_3/r^2\sin^2\theta,$$

for the vectors of the reciprocal basis. The metric coefficients are

$$(g_{ij}) = \begin{pmatrix} 1 & 0 & 0 \\ 0 & r^2 & 0 \\ 0 & 0 & r^2\sin^2\theta \end{pmatrix} \qquad (g^{ij}) = \begin{pmatrix} 1 & 0 & 0 \\ 0 & 1/r^2 & 0 \\ 0 & 0 & 1/r^2\sin^2\theta \end{pmatrix}.$$

Exercise 4.4 The unit basis vectors of spherical coordinates are denoted $\hat{\mathbf{r}}$, $\hat{\boldsymbol{\theta}}$, and $\hat{\boldsymbol{\phi}}$. Give expressions relating the various sets of components (v^1, v^2, v^3), (v_1, v_2, v_3), and (v_r, v_θ, v_ϕ) of a vector $\mathbf{v}$.

We see that both the spherical and cylindrical frames are orthogonal. In this case the reciprocal basis vectors have the same directions as the vectors of the main basis, but have reciprocal lengths so that $|\mathbf{r}_i||\mathbf{r}^i| = 1$ for each i.

In § 2.4 we obtained the transformation laws that apply to the components of a vector under a change of frame. We now extend these laws to the case of a vector field in general coordinates. Because

$$\mathbf{r}_i = \frac{\partial \mathbf{r}}{\partial q^i} = \frac{\partial \mathbf{r}}{\partial \tilde{q}^j}\frac{\partial \tilde{q}^j}{\partial q^i} = \tilde{\mathbf{r}}_j \frac{\partial \tilde{q}^j}{\partial q^i}$$

we can write

$$\mathbf{r}_i = A_i^j \tilde{\mathbf{r}}_j, \qquad A_i^j = \frac{\partial \tilde{q}^j}{\partial q^i}, \tag{4.3}$$

to describe the change of frame. For the inverse transformation

$$\tilde{\mathbf{r}}_i = \tilde{A}_i^j \mathbf{r}_j, \qquad \tilde{A}_i^j = \frac{\partial q^j}{\partial \tilde{q}^i}. \tag{4.4}$$

The results of § 2.4 now apply, and we can write immediately

$$\tilde{f}^i = A_j^i f^j, \qquad \tilde{f}_i = \tilde{A}_i^j f_j, \qquad f^i = \tilde{A}_j^i \tilde{f}^j, \qquad f_i = A_i^j \tilde{f}_j,$$

for the transformation laws pertaining to the components of a vector field $\mathbf{f}$. The f^i are still termed contravariant components of $\mathbf{f}$, while the f_i are still termed covariant components — the main difference is that now A_i^j and $\tilde{A}_i^j$ can change from point to point in space. Note that if all components of a vector vanish in one basis then they do so in every other basis. The transformation laws for the components of higher-rank tensor fields are written similarly.

Exercise 4.5 A set of *oblique rectilinear coordinates* (u, v) in the plane is related to a set of Cartesian coordinates (x, y) by the transformation equations

$$x = u\cos\alpha + v\cos\beta, \qquad y = u\sin\alpha + v\sin\beta,$$

where α, β are constants. (a) Sketch a few u and v coordinate curves superimposed on the xy-plane. (b) Find the basis vectors $\mathbf{r}_i$ $(i = 1, 2)$ and the reciprocal basis vectors $\mathbf{r}^i$ in the (u, v) system. Use these to calculate the metric coefficients. (c) Let $\mathbf{z}$ be a given vector. In the oblique system find the covariant components of $\mathbf{z}$ in terms of the contravariant components of $\mathbf{z}$.

4.2 Differentials and the Nabla Operator

Let us consider the infinitesimal vector from point (q^1, q^2, q^3) to $(q^1 + dq^1, q^2 + dq^2, q^3 + dq^3)$, denoted by $d\mathbf{r}$:

$$d\mathbf{r} = \frac{\partial \mathbf{r}}{\partial q^i} dq^i = \mathbf{r}_i dq^i.$$

Note that by this we can define dq^i as

$$dq^i = \mathbf{r}^i \cdot d\mathbf{r} = d\mathbf{r} \cdot \mathbf{r}^i. \tag{4.5}$$

(Here there is no summation over i. Note also that we only introduce spatial coordinates having superscripts. However, for Cartesian frames it is common to see subscripts used exclusively.) The length of this infinitesimal vector is defined by

$$(ds)^2 = d\mathbf{r} \cdot d\mathbf{r} = \mathbf{r}_i dq^i \cdot \mathbf{r}_j dq^j = g_{ij} dq^i dq^j. \tag{4.6}$$

On the right we have a quadratic form with respect to the variables dq^i; the coefficients of this quadratic form are the covariant components of the metric tensor. In a Cartesian frame (4.6) takes the familiar form

$$(ds)^2 = (dx^1)^2 + (dx^2)^2 + (dx^3)^2.$$

In cylindrical and spherical frames we have

$$(ds)^2 = (d\rho)^2 + \rho^2 (d\phi)^2 + (dz)^2,$$
$$(ds)^2 = (dr)^2 + r^2 (d\theta)^2 + r^2 \sin^2\theta (d\phi)^2,$$

respectively.

Exercise 4.6 Find $(ds)^2$ for the oblique system of Exercise 4.5.

Let $f(q^1, q^2, q^3)$ be a scalar differentiable function of the variables q^i. Its differential is

$$df(q^1, q^2, q^3) = \frac{\partial f(q^1, q^2, q^3)}{\partial q^i} dq^i.$$

Here and in similar situations the i in the denominator stands in the lower position and so we must sum over i. Using (4.5) we can write

$$df = \mathbf{r}^i \frac{\partial f}{\partial q^i} \cdot d\mathbf{r}.$$

The first multiplier on the right is a vector $\mathbf{r}^i \partial f / \partial q^i$. Let us introduce a symbolic vector

$$\nabla = \mathbf{r}^i \frac{\partial}{\partial q^i}$$

called the *nabla operator*, whose action on a function f is as given above:

$$\nabla f = \mathbf{r}^i \frac{\partial f}{\partial q^i}$$

(we repeat that there is summation over i here). The nabla operator is often referred to as the gradient operator.

Exercise 4.7 (a) Show that the unit normal to the surface $\varphi(q^1, q^2, q^3) = c$ = constant is given by

$$\mathbf{n} = \frac{g^{ij}\dfrac{\partial \varphi}{\partial q^i}}{\sqrt{g^{mn}\dfrac{\partial \varphi}{\partial q^m}\dfrac{\partial \varphi}{\partial q^n}}}\mathbf{r}_j.$$

(b) Show that the angle at a point of intersection of the surfaces

$$\varphi(q^1, q^2, q^3) = c_1, \qquad \psi(q^1, q^2, q^3) = c_2,$$

is given by

$$\cos\theta = \frac{g^{ij}\dfrac{\partial \varphi}{\partial q^i}\dfrac{\partial \psi}{\partial q^j}}{\sqrt{g^{mn}\dfrac{\partial \varphi}{\partial q^m}\dfrac{\partial \varphi}{\partial q^n}g^{rt}\dfrac{\partial \psi}{\partial q^r}\dfrac{\partial \psi}{\partial q^t}}}$$

(c) Show that the angle between the coordinate surfaces $q^1 = c_1, q^2 = c_2$ is given by

$$\cos\theta_{12} = \frac{g^{12}}{\sqrt{g^{11}g^{22}}}.$$

(d) Derive the condition for orthogonality of surfaces

$$g^{ij}\frac{\partial \varphi}{\partial q^i}\frac{\partial \psi}{\partial q^j} = 0.$$

Exercise 4.8 Show that the gradient operation in the cylindrical and spherical frames is given by the formulas

$$\nabla f = \hat{\boldsymbol{\rho}}\frac{\partial f}{\partial \rho} + \hat{\boldsymbol{\phi}}\frac{1}{\rho}\frac{\partial f}{\partial \phi} + \hat{\mathbf{z}}\frac{\partial f}{\partial z},$$
$$\nabla f = \hat{\mathbf{r}}\frac{\partial f}{\partial r} + \hat{\boldsymbol{\theta}}\frac{1}{r}\frac{\partial f}{\partial \theta} + \hat{\boldsymbol{\phi}}\frac{1}{r\sin\theta}\frac{\partial f}{\partial \phi},$$

respectively. What are the expressions for these when using the components connected with the triads $\mathbf{r}_i$ and $\mathbf{r}^i$?

Let us find a formula for the differential of a vector function $\mathbf{f}(q^1, q^2, q^3)$:

$$d\mathbf{f}(q^1, q^2, q^3) = \frac{\partial \mathbf{f}(q^1, q^2, q^3)}{\partial q^i}dq^i.$$

We use (4.5) again. Then

$$
\begin{aligned}
d\mathbf{f}(q^1, q^2, q^3) &= \frac{\partial \mathbf{f}(q^1, q^2, q^3)}{\partial q^i}(\mathbf{r}^i \cdot d\mathbf{r}) \\
&= (d\mathbf{r} \cdot \mathbf{r}^i)\frac{\partial \mathbf{f}(q^1, q^2, q^3)}{\partial q^i} \\
&= d\mathbf{r} \cdot \left(\mathbf{r}^i \frac{\partial \mathbf{f}(q^1, q^2, q^3)}{\partial q^i}\right).
\end{aligned}
$$

Thus we can represent this as

$$d\mathbf{f} = d\mathbf{r} \cdot \nabla \mathbf{f}.$$

The quantity $\nabla \mathbf{f}$, known as the *gradient* of $\mathbf{f}$, is clearly a tensor of the second rank. With the aid of transposition we can present $d\mathbf{f}$ in the form

$$d\mathbf{f} = \nabla \mathbf{f}^T \cdot d\mathbf{r}.$$

Sometimes $\nabla \mathbf{f}^T$ is called the gradient of $\mathbf{f}$; it is also called the derivative of $\mathbf{f}$ in the direction of $\mathbf{r}$ and denoted as

$$\frac{d\mathbf{f}}{d\mathbf{r}} = \nabla \mathbf{f}^T.$$

An application of the gradient to a tensor brings a new tensor whose rank is one higher than the rank of the original tensor.

For a pair of vectors we introduced two types of multiplication: the dot and cross product operations. We can apply these to the pair consisting of the nabla operator and a vector function $\mathbf{f} = \mathbf{f}(q^1, q^2, q^3)$. In this way we get two operations: the *divergence*

$$\operatorname{div} \mathbf{f} = \nabla \cdot \mathbf{f} = \mathbf{r}^i \cdot \frac{\partial \mathbf{f}}{\partial q^i},$$

and the *rotation*

$$\operatorname{rot} \mathbf{f} = \nabla \times \mathbf{f} = \mathbf{r}^i \times \frac{\partial \mathbf{f}}{\partial q^i}. \tag{4.7}$$

The vector $\boldsymbol{\omega} = \frac{1}{2} \operatorname{rot} \mathbf{f}$ is called the *curl* of $\mathbf{f}$. In terms of the curl of $\mathbf{f}$ we can introduce a tensor $\boldsymbol{\Omega}$ as

$$\boldsymbol{\Omega} = \mathbf{E} \times \boldsymbol{\omega}.$$

This tensor is antisymmetric (the reader should verify this), and is called the tensor of spin. It can be shown that

$$\boldsymbol{\Omega} = \frac{1}{2}\left(\nabla \mathbf{f}^T - \nabla \mathbf{f}\right).$$

Similarly we can introduce the divergence and rotation operations for a tensor $\mathbf{A}$ of any rank:

$$\nabla \cdot \mathbf{A} = \mathbf{r}^i \cdot \frac{\partial}{\partial q^i}\mathbf{A}, \qquad \nabla \times \mathbf{A} = \mathbf{r}^i \times \frac{\partial}{\partial q^i}\mathbf{A}.$$

Let us see what happens when we apply the nabla operator to the radius vector:

$$\nabla \mathbf{r} = \mathbf{r}^i \frac{\partial}{\partial q^i}\mathbf{r} = \mathbf{r}^i \mathbf{r}_i = \mathbf{E},$$

$$\nabla \cdot \mathbf{r} = \mathbf{r}^i \cdot \frac{\partial}{\partial q^i}\mathbf{r} = \mathbf{r}^i \cdot \mathbf{r}_i = 3,$$

and

$$\nabla \times \mathbf{r} = \mathbf{r}^i \times \frac{\partial}{\partial q^i}\mathbf{r} = \mathbf{r}^i \times \mathbf{r}_i = \mathbf{0}.$$

Exercise 4.9 Calculate $\nabla \mathbf{E}$, $\nabla \cdot \mathbf{E}$, $\nabla \times \mathbf{E}$, $\nabla \boldsymbol{\mathcal{E}}$, $\nabla \cdot \boldsymbol{\mathcal{E}}$, and $\nabla \times \boldsymbol{\mathcal{E}}$.

Exercise 4.10 Let f and g be functions, let $\mathbf{f}$ be a vector, and let $\mathbf{Q}$ be a tensor of the second rank. Verify the following identities:

$$\begin{aligned}
\nabla(fg) &= g\nabla f + f\nabla g, \\
\nabla(f\mathbf{f}) &= (\nabla f)\mathbf{f} + f\nabla \mathbf{f}, \\
\nabla(f\mathbf{Q}) &= (\nabla f)\mathbf{Q} + f\nabla \mathbf{Q}.
\end{aligned}$$

Exercise 4.11 Show that the following identities hold ($\mathbf{f}$ and $\mathbf{g}$ are any vectors):

$$\begin{aligned}
\nabla(\mathbf{f}\cdot\mathbf{g}) &= (\nabla\mathbf{f})\cdot\mathbf{g} + \mathbf{f}\cdot\nabla\mathbf{g}^T = (\nabla\mathbf{f})\cdot\mathbf{g} + (\nabla\mathbf{g})\cdot\mathbf{f}, \\
\nabla(\mathbf{f}\times\mathbf{g}) &= (\nabla\mathbf{f})\times\mathbf{g} - (\nabla\mathbf{g})\times\mathbf{f}, \\
\nabla\times(\mathbf{f}\times\mathbf{g}) &= \mathbf{g}\cdot\nabla\mathbf{f} - \mathbf{g}\nabla\cdot\mathbf{f} - \mathbf{f}\cdot\nabla\mathbf{g} + \mathbf{f}\nabla\cdot\mathbf{g}, \\
\nabla\cdot(\mathbf{f}\mathbf{g}) &= (\nabla\cdot\mathbf{f})\mathbf{g} + \mathbf{f}\nabla\cdot\mathbf{g}, \\
\nabla\cdot(\mathbf{f}\times\mathbf{g}) &= \mathbf{g}\cdot(\nabla\times\mathbf{f}) - \mathbf{f}\cdot(\nabla\times\mathbf{g}).
\end{aligned}$$

4.3 Differentiation of a Vector Function

We have differentiated vector functions with respect to q^i, with the tacit understanding that the formulas necessary to do this resemble those from ordinary calculus. Moreover, the reader certainly knows that to differentiate a vector function $\mathbf{f}(x_1, x_2, x_3) = f^k(x_1, x_2, x_3)\mathbf{i}_k$ with respect to a Cartesian variable x_i it is enough to differentiate each component of $\mathbf{f}$ with respect to this variable:

$$\frac{\partial}{\partial x_i}\mathbf{f}(x_1, x_2, x_3) = \frac{\partial f^k(x_1, x_2, x_3)}{\partial x_i}\mathbf{i}_k. \tag{4.8}$$

Let us consider how to differentiate a vector function $\mathbf{f}$ that is written out in curvilinear coordinates: $\mathbf{f}(q^1, q^2, q^3) = f^i(q^1, q^2, q^3)\mathbf{r}_i$. In this case the frame vectors $\mathbf{r}_i$ depend on q^k as well, which means that simple differentiation of the components of $\mathbf{f}$ does not result in the needed formula. To understand this let us consider a constant function $\mathbf{f}(q^1, q^2, q^3) = \mathbf{c}$. By the general definition of the partial derivative we have $\partial\mathbf{f}(q^1, q^2, q^3)/\partial q^i = \mathbf{0}$. But the components of $\mathbf{f}$ are not constant since the $\mathbf{r}_k$ are variable, hence the derivatives of the components of this function are not zero.

The derivative of the product of a simple function and a vector function is taken by the product rule:

$$\begin{aligned}\frac{\partial}{\partial q^k}\mathbf{f}(q^1, q^2, q^3) &= \frac{\partial}{\partial q^k}\left[f^i(q^1, q^2, q^3)\mathbf{r}_i\right] \\ &= \frac{\partial f^i(q^1, q^2, q^3)}{\partial q^k}\mathbf{r}_i + f^i(q^1, q^2, q^3)\frac{\partial \mathbf{r}_i}{\partial q^k}.\end{aligned}$$

So the derivative of a vector function written in component form consists of two terms. The first is the same as for the derivative in the Cartesian frame (4.8), and the other contains the derivatives of the frame vectors. Thus we need to find the latter.

4.4 Derivatives of the Frame Vectors

We would like to find the value of the derivative

$$\frac{\partial}{\partial q^j}\mathbf{r}_i = \frac{\partial}{\partial q^j}\left(\frac{\partial}{\partial q^i}\mathbf{r}\right) = \frac{\partial^2 \mathbf{r}}{\partial q^j \partial q^i} = \frac{\partial^2 \mathbf{r}}{\partial q^i \partial q^j} = \frac{\partial}{\partial q^i}\mathbf{r}_j. \tag{4.9}$$

Of course if we have the expression for the Cartesian components of the frame vector we can compute these derivatives (and will do so in the exercises); however in the general case we exploit only the fact that the partial derivative of a frame vector is a vector as well, hence it can be expanded in the same frame $\mathbf{r}_i$, $i = 1, 2, 3$. Let us denote the coefficients of the expansion by Γ_{ij}^k:

$$\frac{\partial}{\partial q^j}\mathbf{r}_i = \Gamma_{ij}^k \mathbf{r}_k. \tag{4.10}$$

The Γ_{ij}^k are called *Christoffel coefficients of the second kind*, and are frequently denoted by

$$\Gamma_{ij}^k = \left\{ \begin{matrix} k \\ ij \end{matrix} \right\}. \tag{4.11}$$

By (4.9) there is symmetry in the subscripts of the Christoffel symbols:

$$\Gamma_{ij}^k = \Gamma_{ji}^k. \tag{4.12}$$

Now we can write out the formula for the derivative of a vector function in full:

$$\begin{aligned} \frac{\partial}{\partial q^k}\mathbf{f}(q^1, q^2, q^3) &= \frac{\partial f^i(q^1, q^2, q^3)}{\partial q^k}\mathbf{r}_i + \Gamma_{ki}^j f^i(q^1, q^2, q^3)\mathbf{r}_j \\ &= \frac{\partial f^i(q^1, q^2, q^3)}{\partial q^k}\mathbf{r}_i + \Gamma_{kt}^i f^t(q^1, q^2, q^3)\mathbf{r}_i \\ &= \left(\frac{\partial f^i}{\partial q^k} + \Gamma_{kt}^i f^t\right)\mathbf{r}_i. \end{aligned} \tag{4.13}$$

This is called *covariant differentiation*. The coefficients of $\mathbf{r}_i$ are called the *covariant derivatives* of the contravariant components of $\mathbf{f}$, and are denoted by

$$\nabla_k f^i = \frac{\partial f^i}{\partial q^k} + \Gamma_{kt}^i f^t.$$

Let us discuss some properties of the Christoffel coefficients.

4.5 Christoffel Coefficients and their Properties

In a Cartesian frame the Christoffel symbols are zero, so it is impossible to obtain them for another frame by the usual transformation rules for tensor components. This is easy to understand: the Christoffel symbols depend not only on the frame vectors themselves but on their rates of change from point to point, and these rates do not participate in the transformation rules for tensors. So the Christoffel symbols are not the components of a tensor, despite the appearance of their notation. This is why many authors prefer the notation shown on the right side of equation (4.11).

It is important to have formulas for computing the Christoffel symbols. Let us introduce the notation

$$\mathbf{r}_{ij} = \frac{\partial}{\partial q^j}\mathbf{r}_i = \frac{\partial^2 \mathbf{r}}{\partial q^i \partial q^j}.$$

Relation (4.10) can be written as

$$\mathbf{r}_{ij} = \Gamma^k_{ij}\mathbf{r}_k \tag{4.14}$$

from which it follows that

$$\mathbf{r}_{ij} \cdot \mathbf{r}_t = \Gamma^k_{ij}\mathbf{r}_k \cdot \mathbf{r}_t = \Gamma^k_{ij} g_{kt}.$$

The left-hand side of this can be expressed in terms of the components of the metric tensor. Indeed

$$\frac{\partial}{\partial q^j} g_{it} = \frac{\partial}{\partial q^j}\mathbf{r}_i \cdot \mathbf{r}_t = \mathbf{r}_{ij} \cdot \mathbf{r}_t + \mathbf{r}_i \cdot \mathbf{r}_{tj}. \tag{4.15}$$

Similarly

$$\frac{\partial}{\partial q^t} g_{ji} = \frac{\partial}{\partial q^t}\mathbf{r}_j \cdot \mathbf{r}_i = \mathbf{r}_{jt} \cdot \mathbf{r}_i + \mathbf{r}_j \cdot \mathbf{r}_{it}, \tag{4.16}$$

$$\frac{\partial}{\partial q^i} g_{tj} = \frac{\partial}{\partial q^i}\mathbf{r}_t \cdot \mathbf{r}_j = \mathbf{r}_{ti} \cdot \mathbf{r}_j + \mathbf{r}_t \cdot \mathbf{r}_{ji}. \tag{4.17}$$

Now we obtain $\mathbf{r}_{ij} \cdot \mathbf{r}_t$ by subtracting (4.16) from the sum of (4.15) and (4.17) and dividing this by 2:

$$\mathbf{r}_{ij} \cdot \mathbf{r}_t = \frac{1}{2}\left(\frac{\partial g_{it}}{\partial q^j} + \frac{\partial g_{tj}}{\partial q^i} - \frac{\partial g_{ji}}{\partial q^t}\right) = \Gamma_{ijt}. \tag{4.18}$$

The quantities Γ_{ijk} are called *Christoffel coefficients of the first kind.* They are denoted frequently as $[ij,k] = \Gamma_{ijk}$. It is clear that they are symmetric in the first two subscripts: $\Gamma_{ijk} = \Gamma_{jik}$. Thus we have

$$\Gamma^k_{ij} g_{kt} = \Gamma_{ijt} \tag{4.19}$$

and it follows that

$$\Gamma^k_{ij} = g^{kt}\Gamma_{ijt}. \tag{4.20}$$

Using these we can obtain

$$\frac{\partial \mathbf{r}^j}{\partial q^i} = -\Gamma^j_{it}\mathbf{r}^t. \tag{4.21}$$

Indeed

$$0 = \frac{\partial}{\partial q^t}\delta^i_j = \frac{\partial}{\partial q^t}\left(\mathbf{r}^i \cdot \mathbf{r}_j\right) = \frac{\partial \mathbf{r}^i}{\partial q^t} \cdot \mathbf{r}_j + \mathbf{r}^i \cdot \mathbf{r}_{jt} = \frac{\partial \mathbf{r}^i}{\partial q^t} \cdot \mathbf{r}_j + \mathbf{r}^i \cdot \Gamma^k_{jt}\mathbf{r}_k,$$

so

$$\frac{\partial \mathbf{r}^i}{\partial q^t} \cdot \mathbf{r}_j = -\mathbf{r}^i \cdot \Gamma^k_{jt}\mathbf{r}_k = -\Gamma^k_{jt}\delta^i_k = -\Gamma^i_{jt}$$

and we have

$$\frac{\partial \mathbf{r}^i}{\partial q^t} = -\Gamma^i_{jt}\mathbf{r}^j.$$

It is also useful to have formulas for transformation of the Christoffel symbols under change of coordinates. Suppose the new coordinates $\tilde{q}^i$ are determined by the old coordinates q^i through relations of the form $\tilde{q}^i = \tilde{q}^i(q^1, q^2, q^3)$, and refer back to equations (4.3)–(4.4). Now (4.14) applied in the new system gives

$$\tilde{\Gamma}^k_{ij}\tilde{\mathbf{r}}_k = \tilde{\mathbf{r}}_{ij}. \tag{4.22}$$

But

$$\tilde{\mathbf{r}}_{ij} = \frac{\partial}{\partial \tilde{q}^j}\tilde{\mathbf{r}}_i = \frac{\partial q^m}{\partial \tilde{q}^j}\frac{\partial}{\partial q^m}\tilde{\mathbf{r}}_i = \tilde{A}^m_j \frac{\partial}{\partial q^m}(\tilde{A}^n_i \mathbf{r}_n),$$

and expansion by the product rule gives

$$\begin{aligned}
\tilde{\mathbf{r}}_{ij} &= \tilde{A}^m_j \left(\frac{\partial \tilde{A}^n_i}{\partial q^m}\mathbf{r}_n + \tilde{A}^n_i \frac{\partial \mathbf{r}_n}{\partial q^m}\right) \\
&= \tilde{A}^m_j \frac{\partial \tilde{A}^n_i}{\partial q^m}\mathbf{r}_n + \tilde{A}^m_j \tilde{A}^n_i \Gamma^p_{nm}\mathbf{r}_p \\
&= \tilde{A}^m_j \frac{\partial \tilde{A}^n_i}{\partial q^m} A^k_n \tilde{\mathbf{r}}_k + \tilde{A}^m_j \tilde{A}^n_i \Gamma^p_{nm} A^k_p \tilde{\mathbf{r}}_k.
\end{aligned}$$

Comparison with (4.22) shows that

$$\tilde{\Gamma}^k_{ij} = \tilde{A}^m_j \frac{\partial \tilde{A}^n_i}{\partial q^m} A^k_n + \tilde{A}^m_j \tilde{A}^n_i A^k_p \Gamma^p_{nm}.$$

The presence of the first term on the right means that the Christoffel coefficient is not a tensor of the third rank. This merely confirms our earlier statement of this fact, which was based on different reasoning.

Exercise 4.12 Show that the only nonzero Christoffel coefficients of the first kind for cylindrical coordinates are

$$\Gamma_{221} = -\rho, \qquad \Gamma_{122} = \Gamma_{212} = \rho.$$

Show that the nonzero Christoffel coefficients of the first kind for spherical coordinates are

$$\begin{aligned} \Gamma_{221} &= -r, & \Gamma_{122} &= \Gamma_{212} = r, \\ \Gamma_{331} &= -r\sin^2\theta, & \Gamma_{332} &= -r^2\sin\theta\cos\theta, \\ \Gamma_{313} &= \Gamma_{133} = r\sin^2\theta, & \Gamma_{233} &= \Gamma_{323} = r^2\sin\theta\cos\theta. \end{aligned}$$

Exercise 4.13 Show that the only nonzero Christoffel coefficients of the second kind for cylindrical coordinates are

$$\Gamma^1_{22} = -\rho, \qquad \Gamma^2_{12} = \Gamma^2_{21} = 1/\rho.$$

Show that the nonzero Christoffel coefficients of the second kind for spherical coordinates are

$$\begin{aligned} \Gamma^1_{22} &= -r, & \Gamma^2_{12} &= \Gamma^2_{21} = 1/r, \\ \Gamma^1_{33} &= -r\sin^2\theta, & \Gamma^2_{33} &= -\sin\theta\cos\theta, \\ \Gamma^3_{31} &= \Gamma^3_{13} = 1/r, & \Gamma^3_{23} &= \Gamma^3_{32} = \cot\theta. \end{aligned}$$

Exercise 4.14 A system of plane elliptic coordinates (u, v) is introduced according to the transformation formulas

$$x = c\cosh u\cos v, \qquad y = c\sinh u\sin v,$$

where c is a constant. Find the Christoffel coefficients.

Euclidean vs. non-euclidean spaces

When we derive the length of the elementary vector $d\mathbf{r}$ we write

$$(ds)^2 = d\mathbf{r}\cdot d\mathbf{r} = \mathbf{r}_i dq^i \cdot \mathbf{r}_j dq^j = g_{ij}dq^i dq^j. \tag{4.23}$$

With respect to the variables dq^i and dq^j we have, by construction, a positive definite quadratic form. This is one of the main properties of the metric tensor for a real coordinate space.

If a space is *Euclidean*, that is, if it can be described by a set of Cartesian coordinates x^t, $t = 1, 2, 3$, then the line element $(ds)^2$ can be written as

$$(ds)^2 = \sum_{t=1}^{3} (dx^t)^2. \tag{4.24}$$

Given a set of admissible transformation equations $x^t = x^t(q^n)$, the same line element can be expressed in terms of general coordinates q^n. Since

$$dx^t = \frac{\partial x^t}{\partial q^n} dq^n$$

we have

$$(ds)^2 = \sum_{t=1}^{3} \left(\frac{\partial x^t}{\partial q^n} dq^n \right)^2,$$

and comparison with (4.23) gives

$$g_{ij} = \sum_{t=1}^{3} \frac{\partial x^t}{\partial q^i} \frac{\partial x^t}{\partial q^j}, \qquad i, j = 1, 2, 3, \tag{4.25}$$

for the metric coefficients in the q^n system. We now pose the following question. Suppose a space is originally described in terms of a set of general coordinates q^n. Such a description must include a set of metric coefficients g_{ij} having (4.23) positive definite at each point. Will it be possible to introduce a set of Cartesian coordinates x^t that also describe this space? That is, are we guaranteed the existence of functions $x^t(q^n)$ such that in the resulting coordinates $x^t = x^t(q^n)$ the line element takes the form (4.24)? Put still another way, *is the space Euclidean*? Not necessarily. Given the g_{ij} in the q^n system, (4.25) provides us with six equations for the three unknown functions $x^t(q^n)$. This implies that some additional conditions must be fulfilled by the given metric coefficients. It is possible to formulate these restrictions neatly in terms of a certain fourth rank tensor $R^p_{\cdot ijk}$ as the equality

$$R^p_{\cdot ijk} = 0. \tag{4.26}$$

The tensor $R^p_{\cdot ijk}$ is known as the *Riemann–Christoffel tensor*, and its associated tensor $R_{nijk} = g_{np} R^p_{\cdot ijk}$ is called the *curvature tensor* for the space. It can be shown that

$$R^p_{\cdot ijk} = \frac{\partial \Gamma^p_{ij}}{\partial q^k} - \frac{\partial \Gamma^p_{ik}}{\partial q^j} - \left(\Gamma^p_{mj} \Gamma^m_{ik} - \Gamma^p_{mk} \Gamma^m_{ij} \right).$$

It can also be shown that $R^{p}_{\cdot ijk}$ actually has only six independent components, so (4.26) represents six conditions on the g_{ij} that must hold for the space described by the g_{ij} to be Euclidean.

We have included this information only for the reader's general background, as such considerations turn out to be very important in certain application areas. Further details and more rigorous formulations can be found in some of the more comprehensive references.

4.6 Covariant Differentiation

Let us return to the problem of differentiation of a vector function. We derived the formula for the derivative (4.13)

$$\frac{\partial}{\partial q^k}\mathbf{f} = \frac{\partial}{\partial q^k}(f^i \mathbf{r}_i) = \left(\frac{\partial f^i}{\partial q^k} + \Gamma^i_{kt} f^t\right)\mathbf{r}_i$$

and noted that the coefficients of this expansion denoted by

$$\nabla_k f^i = \frac{\partial f^i}{\partial q^k} + \Gamma^i_{kt} f^t \tag{4.27}$$

are called the covariant derivatives of contravariant components of vector function $\mathbf{f}$. Let us express the same derivatives in terms of the covariant components:

$$\frac{\partial}{\partial q^k}\mathbf{f} = \frac{\partial}{\partial q^k}(f_i \mathbf{r}^i) = \frac{\partial f_i}{\partial q^k}\mathbf{r}^i + f_i \frac{\partial}{\partial q^k}\mathbf{r}^i.$$

Using (4.21) we get

$$\frac{\partial}{\partial q^k}\mathbf{f} = \frac{\partial f_i}{\partial q^k}\mathbf{r}^i - f_i \Gamma^i_{kt}\mathbf{r}^t = \left(\frac{\partial f_i}{\partial q^k} - \Gamma^j_{ki} f_j\right)\mathbf{r}^i.$$

The coefficients of this expansion are called covariant derivatives of the covariant components and are denoted by

$$\nabla_k f_i = \frac{\partial f_i}{\partial q^k} - \Gamma^j_{ki} f_j. \tag{4.28}$$

In these notations we can write out the formula for differentiation in the form

$$\frac{\partial \mathbf{f}}{\partial q^i} = \mathbf{r}^k \nabla_i f_k = \mathbf{r}_j \nabla_i f^j. \tag{4.29}$$

Because

$$\nabla \mathbf{f} = \mathbf{r}^i \frac{\partial}{\partial q^i}(f_j \mathbf{r}^j) = \mathbf{r}^i (\nabla_i f_j) \mathbf{r}^j = \mathbf{r}^i \mathbf{r}^j \nabla_i f_j,$$

we see that $\nabla_i f_j$ is a covariant component of $\nabla \mathbf{f}$. Similarly,

$$\nabla \mathbf{f} = \mathbf{r}^i \frac{\partial}{\partial q^i}(f^j \mathbf{r}_j) = \mathbf{r}^i \mathbf{r}_j \nabla_i f^j$$

shows that $\nabla_i f^j$ is a mixed component of $\nabla \mathbf{f}$.

Exercise 4.15 Write out the covariant derivatives of the contravariant components of a vector $\mathbf{f}$ in plane polar coordinates.

Quick summary

The formulas for differentiation of a vector are

$$\frac{\partial \mathbf{f}}{\partial q^i} = \mathbf{r}^k \nabla_i f_k = \mathbf{r}_j \nabla_i f^j$$

where

$$\nabla_k f_i = \frac{\partial f_i}{\partial q^k} - \Gamma^j_{ki} f_j, \qquad \nabla_k f^i = \frac{\partial f^i}{\partial q^k} + \Gamma^i_{kt} f^t.$$

The formulas

$$\nabla \mathbf{f} = \mathbf{r}^i \mathbf{r}^j \nabla_i f_j = \mathbf{r}^i \mathbf{r}_j \nabla_i f^j$$

show that $\nabla_i f_j$ is a covariant component of $\nabla \mathbf{f}$, while $\nabla_i f^j$ is a mixed component of $\nabla \mathbf{f}$.

4.7 Covariant Derivative of a Second Rank Tensor

Let us find a partial derivative of a tensor $\mathbf{A}$ of the second rank:

$$\begin{aligned}
\frac{\partial}{\partial q^k}\mathbf{A} &= \frac{\partial}{\partial q^k}(a^{ij}\mathbf{r}_i \mathbf{r}_j) \\
&= \frac{\partial a^{ij}}{\partial q^k}\mathbf{r}_i \mathbf{r}_j + a^{ij}(\mathbf{r}_{ik}\mathbf{r}_j + \mathbf{r}_i \mathbf{r}_{jk}) \\
&= \frac{\partial a^{ij}}{\partial q^k}\mathbf{r}_i \mathbf{r}_j + a^{ij}(\Gamma^t_{ik}\mathbf{r}_t \mathbf{r}_j + \mathbf{r}_i \Gamma^t_{jk}\mathbf{r}_t).
\end{aligned}$$

Changing dummy indices we get

$$\frac{\partial}{\partial q^k}\mathbf{A} = \left(\frac{\partial a^{ij}}{\partial q^k} + \Gamma^i_{ks}a^{sj} + \Gamma^j_{ks}a^{is}\right)\mathbf{r}_i\mathbf{r}_j.$$

The parenthetical expression is designated as a covariant derivative:

$$\nabla_k a^{ij} = \frac{\partial a^{ij}}{\partial q^k} + \Gamma^i_{ks}a^{sj} + \Gamma^j_{ks}a^{is}.$$

Similarly

$$\begin{aligned}\frac{\partial}{\partial q^k}\mathbf{A} &= \frac{\partial}{\partial q^k}(a_{ij}\mathbf{r}^i\mathbf{r}^j)\\ &= \frac{\partial a_{ij}}{\partial q^k}\mathbf{r}^i\mathbf{r}^j + a_{ij}\left(\frac{\partial \mathbf{r}^i}{\partial q^k}\mathbf{r}^j + \mathbf{r}^i\frac{\partial \mathbf{r}^j}{\partial q^k}\right)\\ &= \frac{\partial a_{ij}}{\partial q^k}\mathbf{r}^i\mathbf{r}^j - a_{ij}(\Gamma^i_{kt}\mathbf{r}^t\mathbf{r}^j + \mathbf{r}^i\Gamma^j_{kt}\mathbf{r}^t)\\ &= \left(\frac{\partial a_{ij}}{\partial q^k} - \Gamma^s_{ki}a_{sj} - \Gamma^s_{kj}a_{is}\right)\mathbf{r}^i\mathbf{r}^j.\end{aligned}$$

As before we denote

$$\nabla_k a_{ij} = \frac{\partial a_{ij}}{\partial q^k} - \Gamma^s_{ki}a_{sj} - \Gamma^s_{kj}a_{is}.$$

Exercise 4.16 Show that

$$\frac{\partial}{\partial q^k}\mathbf{A} = \left(\frac{\partial a_i^{\cdot j}}{\partial q^k} - \Gamma^s_{ki}a_s^{\cdot j} + \Gamma^j_{ks}a_i^{\cdot s}\right)\mathbf{r}^i\mathbf{r}_j,$$

$$\frac{\partial}{\partial q^k}\mathbf{A} = \left(\frac{\partial a^i_{\cdot j}}{\partial q^k} + \Gamma^i_{ks}a^s_{\cdot j} - \Gamma^s_{kj}a^i_{\cdot s}\right)\mathbf{r}_i\mathbf{r}^j.$$

The expressions in parentheses are all denoted the same way:

$$\nabla_k a_i^{\cdot j} = \frac{\partial a_i^{\cdot j}}{\partial q^k} - \Gamma^s_{ki}a_s^{\cdot j} + \Gamma^j_{ks}a_i^{\cdot s},$$

$$\nabla_k a^i_{\cdot j} = \frac{\partial a^i_{\cdot j}}{\partial q^k} + \Gamma^i_{ks}a^s_{\cdot j} - \Gamma^s_{kj}a^i_{\cdot s}.$$

Quick summary

For a tensor $\mathbf{A}$ of the second rank we have

$$\frac{\partial}{\partial q^k}\mathbf{A} = \nabla_k a^{ij}\mathbf{r}_i\mathbf{r}_j = \nabla_k a_{ij}\mathbf{r}^i\mathbf{r}^j = \nabla_k a_i^{\cdot j}\mathbf{r}^i\mathbf{r}_j = \nabla_k a^i_{\cdot j}\mathbf{r}_i\mathbf{r}^j$$

where

$$\nabla_k a^{ij} = \frac{\partial a^{ij}}{\partial q^k} + \Gamma^i_{ks} a^{sj} + \Gamma^j_{ks} a^{is}, \quad \nabla_k a_{ij} = \frac{\partial a_{ij}}{\partial q^k} - \Gamma^s_{ki} a_{sj} - \Gamma^s_{kj} a_{is},$$

$$\nabla_k a_i^{\cdot j} = \frac{\partial a_i^{\cdot j}}{\partial q^k} - \Gamma^s_{ki} a_s^{\cdot j} + \Gamma^j_{ks} a_i^{\cdot s}, \quad \nabla_k a^i_{\cdot j} = \frac{\partial a^i_{\cdot j}}{\partial q^k} + \Gamma^i_{ks} a^s_{\cdot j} - \Gamma^s_{kj} a^i_{\cdot s}.$$

Exercise 4.17 Show that any of the covariant derivatives of any component of the metric tensor is equal to zero.

Exercise 4.18 Demonstrate that the components of the metric tensor behave as constants under covariant differentiation of components:

$$\nabla_k g^{st} a_t = g^{st} \nabla_k a_t, \qquad \nabla_k g_{st} a^t = g_{st} \nabla_k a^t.$$

4.8 Differential Operations

Here we look further at various differential operations that may be performed on vector and tensor fields. We begin with the rotation of a vector. By (4.7) and (4.29) we have

$$\nabla \times \mathbf{f} = \mathbf{r}^i \times \frac{\partial \mathbf{f}}{\partial q^i} = \mathbf{r}^i \times \mathbf{r}^j \nabla_i f_j = \epsilon^{ijk} \mathbf{r}_k \nabla_i f_j.$$

The use of (4.28) allows us to write this as

$$\nabla \times \mathbf{f} = \epsilon^{ijk} \mathbf{r}_k \left(\frac{\partial f_j}{\partial q^i} - \Gamma^n_{ij} f_n \right). \tag{4.30}$$

Considering the second term in parentheses we note that $\epsilon^{ijk}\Gamma^n_{ij} = \epsilon^{ijk}\Gamma^n_{ji} = \epsilon^{jik}\Gamma^n_{ij}$ by (4.12) and a subsequent renaming of dummy indices. But $\epsilon^{jik} = -\epsilon^{ijk}$, hence $\epsilon^{ijk}\Gamma^n_{ij} = -\epsilon^{ijk}\Gamma^n_{ij}$ so that $\epsilon^{ijk}\Gamma^n_{ij} = 0$. Equation (4.30) is thereby reduced to

$$\nabla \times \mathbf{f} = \mathbf{r}_k \epsilon^{ijk} \frac{\partial f_j}{\partial q^i}.$$

As an example we recall that the vector field $\mathbf{E}$ of electrostatics satisfies $\nabla \times \mathbf{E} = 0$. Such a field is said to be *irrotational.* So the condition for any vector field $\mathbf{f}$ to be irrotational can be written in generalized coordinates as

$$\epsilon^{ijk} \frac{\partial f_j}{\partial q^i} = 0.$$

Of course, the operation $\nabla \times \mathbf{f}$ is given in a Cartesian frame by the familiar formula

$$\nabla \times \mathbf{f} = \begin{vmatrix} \hat{\mathbf{x}} & \hat{\mathbf{y}} & \hat{\mathbf{z}} \\ \dfrac{\partial}{\partial x} & \dfrac{\partial}{\partial y} & \dfrac{\partial}{\partial z} \\ f_x & f_y & f_z \end{vmatrix}.$$

Let us turn to the divergence of $\mathbf{f}$. We start by writing

$$\nabla \cdot \mathbf{f} = \mathbf{r}^i \cdot \frac{\partial \mathbf{f}}{\partial q^i} = \mathbf{r}^i \cdot \mathbf{r}^j \nabla_i f_j = g^{ij} \nabla_i f_j.$$

By the result of Exercise 4.7.3 and equation (4.27) this can be written as

$$\nabla \cdot \mathbf{f} = \nabla_i g^{ij} f_j = \nabla_i f^i = \frac{\partial f^i}{\partial q^i} + \Gamma^i_{in} f^n. \tag{4.31}$$

As was the case with (4.30) this can be simplified; we must first develop a useful identity for Γ^i_{in}. This is done as follows. We begin by writing

$$\frac{\partial \sqrt{g}}{\partial q^n} = \frac{\partial}{\partial q^n}[\mathbf{r}_1 \cdot (\mathbf{r}_2 \times \mathbf{r}_3)] = \frac{\partial \mathbf{r}_1}{\partial q^n} \cdot (\mathbf{r}_2 \times \mathbf{r}_3) + \mathbf{r}_1 \cdot \frac{\partial}{\partial q^n}(\mathbf{r}_2 \times \mathbf{r}_3)$$

where

$$\begin{aligned} \mathbf{r}_1 \cdot \frac{\partial}{\partial q^n}(\mathbf{r}_2 \times \mathbf{r}_3) &= \mathbf{r}_1 \cdot \mathbf{r}_2 \times \frac{\partial \mathbf{r}_3}{\partial q^n} + \mathbf{r}_1 \cdot \frac{\partial \mathbf{r}_2}{\partial q^n} \times \mathbf{r}_3 \\ &= \frac{\partial \mathbf{r}_3}{\partial q^n} \cdot (\mathbf{r}_1 \times \mathbf{r}_2) + \frac{\partial \mathbf{r}_2}{\partial q^n} \cdot (\mathbf{r}_3 \times \mathbf{r}_1) \end{aligned}$$

so that

$$\frac{\partial \sqrt{g}}{\partial q^n} = \frac{\partial \mathbf{r}_1}{\partial q^n} \cdot (\mathbf{r}_2 \times \mathbf{r}_3) + \frac{\partial \mathbf{r}_2}{\partial q^n} \cdot (\mathbf{r}_3 \times \mathbf{r}_1) + \frac{\partial \mathbf{r}_3}{\partial q^n} \cdot (\mathbf{r}_1 \times \mathbf{r}_2).$$

Continuing to rewrite this we have

$$\begin{aligned} \frac{\partial \sqrt{g}}{\partial q^n} &= \mathbf{r}_{1n} \cdot (\mathbf{r}_2 \times \mathbf{r}_3) + \mathbf{r}_{2n} \cdot (\mathbf{r}_3 \times \mathbf{r}_1) + \mathbf{r}_{3n} \cdot (\mathbf{r}_1 \times \mathbf{r}_2) \\ &= \Gamma^i_{1n}\mathbf{r}_i \cdot (\mathbf{r}_2 \times \mathbf{r}_3) + \Gamma^i_{2n}\mathbf{r}_i \cdot (\mathbf{r}_3 \times \mathbf{r}_1) + \Gamma^i_{3n}\mathbf{r}_i \cdot (\mathbf{r}_1 \times \mathbf{r}_2) \\ &= \Gamma^1_{1n}\sqrt{g} + \Gamma^2_{2n}\sqrt{g} + \Gamma^3_{3n}\sqrt{g} \\ &= \sqrt{g}\,\Gamma^i_{in}. \end{aligned}$$

Hence

$$\Gamma^i_{in} = \frac{1}{\sqrt{g}} \frac{\partial \sqrt{g}}{\partial q^n}.$$

This is the needed identity, and with it (4.31) may be written as

$$\nabla \cdot \mathbf{f} = \frac{1}{\sqrt{g}} \frac{\partial}{\partial q^i} \left(\sqrt{g} f^i \right). \tag{4.32}$$

We know that the condition of incompressibility of a liquid in hydromechanics is expressed as

$$\nabla \cdot \mathbf{v} = 0$$

where $\mathbf{v}$ is the velocity of a material point, so in general coordinates we can write it out as

$$\frac{1}{\sqrt{g}} \frac{\partial}{\partial q^i} \left(\sqrt{g} v^i \right) = 0.$$

In electromagnetic theory the magnetic source law states that $\nabla \cdot \mathbf{B} = 0$ where $\mathbf{B}$ is the vector field known as magnetic flux density. In general, a vector field $\mathbf{f}$ is called *solenoidal* if $\nabla \cdot \mathbf{f} = 0$. Of course, $\nabla \cdot \mathbf{f}$ is given in Cartesian frames by the familiar expression

$$\nabla \cdot \mathbf{f} = \frac{\partial f_x}{\partial x} + \frac{\partial f_y}{\partial y} + \frac{\partial f_z}{\partial z}.$$

Exercise 4.19 Use (4.32) to express $\nabla \cdot \mathbf{f}$ in the cylindrical and spherical coordinate systems.

The curl of a tensor field $\mathbf{A}$ may be computed as follows. We start with

$$\begin{aligned}
\nabla \times \mathbf{A} &= \mathbf{r}^k \times \frac{\partial}{\partial q^k} \mathbf{A} = \mathbf{r}^k \times \mathbf{r}^i \mathbf{r}^j \nabla_k a_{ij} = \epsilon^{kin} \mathbf{r}_n \mathbf{r}^j \nabla_k a_{ij} \\
&= \epsilon^{kin} \mathbf{r}_n \mathbf{r}^j \left(\frac{\partial a_{ij}}{\partial q^k} - \Gamma^s_{ki} a_{sj} - \Gamma^s_{kj} a_{is} \right).
\end{aligned}$$

We then use

$$\epsilon^{kin} \Gamma^s_{ki} = 0, \qquad -\mathbf{r}^j \Gamma^s_{kj} = \frac{\partial \mathbf{r}^s}{\partial q^k},$$

to get

$$\nabla \times \mathbf{A} = \epsilon^{kin} \mathbf{r}_n \left(\mathbf{r}^j \frac{\partial a_{ij}}{\partial q^k} + \frac{\partial \mathbf{r}^s}{\partial q^k} a_{is} \right)$$

and hence

$$\nabla \times \mathbf{A} = \epsilon^{kin} \mathbf{r}_n \frac{\partial}{\partial q^k} \left(\mathbf{r}^j a_{ij} \right).$$

For the divergence of a tensor field $\mathbf{A}$, we write

$$\nabla \cdot \mathbf{A} = \mathbf{r}^k \cdot \frac{\partial}{\partial q^k}\mathbf{A} = \mathbf{r}^k \cdot \mathbf{r}_i \mathbf{r}_j \nabla_k a^{ij} = \mathbf{r}_j \nabla_i a^{ij}$$
$$= \mathbf{r}_j \left(\frac{\partial a^{ij}}{\partial q^i} + \Gamma^i_{is} a^{sj} + \Gamma^j_{is} a^{is} \right).$$

But

$$\mathbf{r}_j \Gamma^j_{is} = \frac{\partial \mathbf{r}_s}{\partial q^i}$$

so we have

$$\nabla \cdot \mathbf{A} = \mathbf{r}_j \frac{\partial a^{ij}}{\partial q^i} + \mathbf{r}_j \frac{1}{\sqrt{g}} \frac{\partial \sqrt{g}}{\partial q^s} a^{sj} + \frac{\partial \mathbf{r}_s}{\partial q^i} a^{is}$$
$$= \frac{\partial}{\partial q^i}\left(\mathbf{r}_j a^{ij}\right) + \mathbf{r}_j \frac{1}{\sqrt{g}} \frac{\partial \sqrt{g}}{\partial q^i} a^{ij}.$$

Finally then,

$$\nabla \cdot \mathbf{A} = \frac{1}{\sqrt{g}} \frac{\partial}{\partial q^i}\left(\sqrt{g} a^{ij} \mathbf{r}_j\right).$$

Many problems of mathematical physics reduce to either Poisson's equation or Laplace's equation. The unknown function in these equations can be a scalar function or (as in electrodynamics) a vector function. The equations of the linear theory of elasticity in displacements also contain the Laplacian operator and another type of operation involving the nabla operator. In Cartesian frames it is simple to write out corresponding expressions. Solving corresponding problems with the use of curvilinear coordinates, we need to find the representation of these formulas. We begin with the formulas that relate to the tensor of second rank $\nabla\nabla f$. We have

$$\nabla\nabla f = \mathbf{r}^i \frac{\partial}{\partial q^i}\left(\mathbf{r}^j \frac{\partial}{\partial q^j} f\right) = \mathbf{r}^i \left(\frac{\partial}{\partial q^i} \frac{\partial f}{\partial q^j} - \Gamma^k_{ij} \frac{\partial f}{\partial q^k} \right) \mathbf{r}^j,$$

hence

$$\nabla\nabla f = \mathbf{r}^i \mathbf{r}^j \left(\frac{\partial^2 f}{\partial q^i \partial q^j} - \Gamma^k_{ij} \frac{\partial f}{\partial q^k} \right). \tag{4.33}$$

From this we see that

$$(\nabla\nabla f)^T = \nabla\nabla f \tag{4.34}$$

which is obvious by symmetry (in i and j) of the Christoffel coefficient and the rest of the expression on the right side of (4.33). A formal insertion[2] of the dot product operation between the vectors of (4.33) allows us to generate an expression for the Laplacian $\nabla^2 f \equiv \nabla \cdot \nabla f$:

$$\nabla^2 f = \mathbf{r}^i \cdot \mathbf{r}^j \left(\frac{\partial^2 f}{\partial q^i \partial q^j} - \Gamma_{ij}^k \frac{\partial f}{\partial q^k} \right) = g^{ij} \left(\frac{\partial^2 f}{\partial q^i \partial q^j} - \Gamma_{ij}^k \frac{\partial f}{\partial q^k} \right). \quad (4.35)$$

Hence Laplace's equation $\nabla^2 f = 0$ appears in generalized coordinates as

$$g^{ij} \left(\frac{\partial^2 f}{\partial q^i \partial q^j} - \Gamma_{ij}^k \frac{\partial f}{\partial q^k} \right) = 0.$$

In a Cartesian frame (4.35) gives us

$$\nabla^2 f = \frac{\partial^2 f}{\partial x^2} + \frac{\partial^2 f}{\partial y^2} + \frac{\partial^2 f}{\partial z^2},$$

while in cylindrical and spherical frames

$$\nabla^2 f = \frac{1}{\rho} \frac{\partial}{\partial \rho} \left(\rho \frac{\partial f}{\partial \rho} \right) + \frac{1}{\rho^2} \frac{\partial^2 f}{\partial \phi^2} + \frac{\partial^2 f}{\partial z^2},$$

$$\nabla^2 f = \frac{1}{r^2} \frac{\partial}{\partial r} \left(r^2 \frac{\partial f}{\partial r} \right) + \frac{1}{r^2 \sin\theta} \frac{\partial}{\partial \theta} \left(\sin\theta \frac{\partial f}{\partial \theta} \right) + \frac{1}{r^2 \sin^2\theta} \frac{\partial^2 f}{\partial \phi^2},$$

respectively.

Exercise 4.20 Show that a formal insertion of the cross product operation between the vectors of (4.33) leads to the useful identity

$$\nabla \times \nabla f = 0,$$

holding for any scalar field f.

Since usual and covariant differentiation amount to the same thing for a scalar f, we can write

$$\nabla \nabla f = \mathbf{r}^i \frac{\partial}{\partial q^i} \mathbf{r}^j \frac{\partial}{\partial q^j} f = \mathbf{r}^i \frac{\partial}{\partial q^i} \left(\mathbf{r}^j \nabla_j f \right) = \mathbf{r}^i \mathbf{r}^j \nabla_i \nabla_j f = \mathbf{r}^j \mathbf{r}^i \nabla_i \nabla_j f$$

where in the last step we used (4.34). The corresponding result for the Laplacian is

$$\nabla^2 f = \mathbf{r}^i \cdot \mathbf{r}^j \nabla_i \nabla_j f = g^{ij} \nabla_i \nabla_j f = \nabla^j \nabla_j f$$

[2] This sort of operation can be done with any dyad of vectors; we can generate a scalar $\mathbf{a} \cdot \mathbf{b}$ from $\mathbf{ab}$ by inserting a dot. This operation can have additional meaning when done with a tensor $\mathbf{A}$: if we write the tensor in mixed form $a_i^{\cdot j} \mathbf{r}^i \mathbf{r}_j$ we obtain $a_i^{\cdot j} \mathbf{r}^i \cdot \mathbf{r}_j = a_i^{\cdot i}$. This is the first invariant of $\mathbf{A}$, known as the trace of $\mathbf{A}$.

where

$$\nabla^j \equiv g^{ij}\nabla_i.$$

The Laplacian of a vector arises in physical applications such as electromagnetic field theory. For this we have

$$\nabla^2\mathbf{f} = \nabla\cdot\nabla\mathbf{f} = \mathbf{r}^k\cdot\frac{\partial}{\partial q^k}\mathbf{r}^i\mathbf{r}_j\nabla_i f^j = \mathbf{r}^k\cdot\mathbf{r}^i\mathbf{r}_j\nabla_k\nabla_i f^j = g^{ki}\mathbf{r}_j\nabla_k\nabla_i f^j$$

hence

$$\nabla^2\mathbf{f} = \mathbf{r}_j\nabla^i\nabla_i f^j.$$

Also

$$\nabla\nabla\cdot\mathbf{f} = \mathbf{r}^i\frac{\partial}{\partial q^i}\left[\frac{1}{\sqrt{g}}\frac{\partial}{\partial q^j}\left(\sqrt{g}f^j\right)\right] = \mathbf{r}^i\nabla_i\nabla_j f^j.$$

It is possible to demonstrate that

$$\nabla\times\nabla\times\mathbf{f} = \nabla\nabla\cdot\mathbf{f} - \nabla^2\mathbf{f}.$$

4.9 Orthogonal Coordinate Systems

The most frequently used coordinate frames are Cartesian, cylindrical, and spherical. All of these are orthogonal. There are many other orthogonal coordinate frames in use as well. It is sensible to give a general treatment of these systems because mutual orthogonality of the frame vectors leads to simplification in many formulae. Additional motivation is provided by the fact that in applications it is important to know the magnitudes of field components. The general formulations given above are not very convenient for this; the general frame vectors are not of unit length, hence the magnitude of the projection of a vector onto an orthogonal frame direction is not the corresponding component of the vector. The physical components of a vector are conveniently displayed in orthogonal frames with use of the so-called Lamé coefficients.

In this section then, we consider frames where the coordinate vectors are mutually orthogonal:

$$\mathbf{r}_i\cdot\mathbf{r}_j = 0, \qquad i\neq j.$$

The Lamé coefficients H_i are

$$(H_i)^2 = g_{ii} = \mathbf{r}_i\cdot\mathbf{r}_i, \qquad i = 1,2,3.$$

We note that H_i is the length of the frame vector $\mathbf{r}_i$. At each (q^1, q^2, q^3) the coordinate frame is orthogonal. In the orthogonal frame, by the construction of the reciprocal basis, the vectors $\mathbf{r}^i$ are co-directed with $\mathbf{r}_i$ and the product of their lengths is 1. Thus we have

$$\mathbf{r}^i = \mathbf{r}_i/(H_i)^2, \qquad i = 1, 2, 3.$$

Let us introduce the frame whose vectors are co-directed with the basis vectors but have unit length:

$$\hat{\mathbf{r}}_i = \mathbf{r}_i/H_i, \quad i = 1, 2, 3.$$

Thus at each point the vectors $\hat{\mathbf{r}}_i$ form a Cartesian basis, a frame that rotates when the origin of the frame moves from point to point. We shall present all the main formulas of differentiation when vectors are given in this basis. They are not convenient in theory since many useful properties of symmetry are lost, but they are necessary when doing calculations in corresponding coordinates. We begin by noting that

$$\mathbf{r}_i = H_i \hat{\mathbf{r}}_i, \qquad \mathbf{r}^i = \hat{\mathbf{r}}_i/H_i, \qquad i = 1, 2, 3.$$

Let us compute the $\hat{\mathbf{r}}_i$ in the cylindrical and spherical systems. In a cylindrical frame where

$$H_1 = 1, \qquad H_2 = \rho, \qquad H_3 = 1,$$

we have

$$\begin{aligned}
\hat{\mathbf{r}}_1 &= \hat{\mathbf{x}} \cos\phi + \hat{\mathbf{y}} \sin\phi, \\
\hat{\mathbf{r}}_2 &= -\hat{\mathbf{x}} \sin\phi + \hat{\mathbf{y}} \cos\phi, \\
\hat{\mathbf{r}}_3 &= \hat{\mathbf{z}}.
\end{aligned}$$

In a spherical frame where

$$H_1 = 1, \qquad H_2 = r, \qquad H_3 = r\sin\theta,$$

we have

$$\begin{aligned}
\hat{\mathbf{r}}_1 &= \hat{\mathbf{x}} \sin\theta\cos\phi + \hat{\mathbf{y}} \sin\theta\sin\phi + \hat{\mathbf{z}}\cos\theta, \\
\hat{\mathbf{r}}_2 &= \hat{\mathbf{x}} \cos\theta\cos\phi + \hat{\mathbf{y}} \cos\theta\sin\phi - \hat{\mathbf{z}}\sin\theta, \\
\hat{\mathbf{r}}_3 &= -\hat{\mathbf{x}} \sin\phi + \hat{\mathbf{y}} \cos\phi.
\end{aligned}$$

Differentiation in the orthogonal basis

Using the definition we can represent the ∇-operator in new terms as

$$\nabla = \frac{\hat{\mathbf{r}}_i}{H_i}\frac{\partial}{\partial q^i}.$$

(In this formula there *is* summation over i, and this continues to hold in all formulas of this section below. The convention on summation over sub- and super-indices is modified here since in a Cartesian system, which the frame $\hat{\mathbf{r}}_i$ locally constitutes, the reciprocal and main bases coincide.) Now let us find the formulas of differentiation of the new frame vectors. We begin with the formula (4.18):

$$\begin{aligned}\mathbf{r}_{ij}\cdot\mathbf{r}_t &= \frac{1}{2}\left(\frac{\partial g_{it}}{\partial q^j} + \frac{\partial g_{tj}}{\partial q^i} - \frac{\partial g_{ji}}{\partial q^t}\right)\\ &= H_t\frac{\partial H_i}{\partial q^j}\delta_{it} + H_t\frac{\partial H_j}{\partial q^i}\delta_{jt} - H_i\frac{\partial H_j}{\partial q^t}\delta_{ij}.\end{aligned}$$

Here we used the fact that $g_{ij} = 0$ if $i \neq j$. On the other hand

$$\begin{aligned}\mathbf{r}_{ij}\cdot\mathbf{r}_t &= \frac{\partial}{\partial q^j}(H_i\hat{\mathbf{r}}_i)\cdot H_t\hat{\mathbf{r}}_t\\ &= \frac{\partial H_i}{\partial q^j}\hat{\mathbf{r}}_i\cdot H_t\hat{\mathbf{r}}_t + H_i\frac{\partial \hat{\mathbf{r}}_i}{\partial q^j}\cdot H_t\hat{\mathbf{r}}_t\\ &= \frac{\partial H_i}{\partial q^j}H_t\delta_{it} + H_iH_t\frac{\partial \hat{\mathbf{r}}_i}{\partial q^j}\cdot\hat{\mathbf{r}}_t.\end{aligned}$$

Thus

$$\frac{\partial \hat{\mathbf{r}}_i}{\partial q^j}\cdot\hat{\mathbf{r}}_t = \frac{1}{H_i}\frac{\partial H_t}{\partial q^i}\delta_{jt} - \frac{1}{H_t}\frac{\partial H_i}{\partial q^t}\delta_{ij}.$$

Therefore

$$\frac{\partial \hat{\mathbf{r}}_i}{\partial q^j} = \sum_{t=1}^{3}\left(\frac{1}{H_i}\frac{\partial H_t}{\partial q^i}\delta_{jt} - \frac{1}{H_t}\frac{\partial H_i}{\partial q^t}\delta_{ij}\right)\hat{\mathbf{r}}_t.$$

(Note that the components of vectors are given in the frame of the same $\hat{\mathbf{r}}_i$.) Using this we derive

$$\nabla\mathbf{f} = \sum_{i,j=1}^{3}\frac{\hat{\mathbf{r}}_j}{H_j}\frac{\partial}{\partial q^j}\left(f_i\hat{\mathbf{r}}_i\right).$$

The gradient, divergence, and rotation of a vector field $\mathbf{f}$ are given by the equations

$$\nabla \mathbf{f} = \hat{\mathbf{r}}_i \hat{\mathbf{r}}_j \left(\frac{1}{H_i} \frac{\partial f_j}{\partial q^i} - \frac{f_i}{H_i H_j} \frac{\partial H_i}{\partial q^j} + \delta_{ij} \frac{f_k}{H_k} \frac{1}{H_i} \frac{\partial H_i}{\partial q^k} \right)$$

$$\nabla \cdot \mathbf{f} = \frac{1}{H_1 H_2 H_3} \left(\frac{\partial}{\partial q^1} (H_2 H_3 f_1) + \frac{\partial}{\partial q^2} (H_3 H_1 f_2) + \frac{\partial}{\partial q^3} (H_1 H_2 f_3) \right) \tag{4.36}$$

and

$$\nabla \times \mathbf{f} = \frac{1}{2} \frac{\hat{\mathbf{r}}_i \times \hat{\mathbf{r}}_j}{H_i H_j} \left(\frac{\partial}{\partial q^i} (H_j f_j) - \frac{\partial}{\partial q^j} (H_i f_i) \right) \tag{4.37}$$

Using the gradient we can write out the tensor of small strains for a displacement vector $\mathbf{u} = u_i \hat{\mathbf{r}}_i$, which is the main object of linear elasticity:

$$\begin{aligned} \boldsymbol{\epsilon} &= \frac{1}{2} \left(\nabla \mathbf{u} + \nabla \mathbf{u}^T \right) \\ &= \frac{1}{2} \hat{\mathbf{r}}_i \hat{\mathbf{r}}_j \left(\frac{1}{H_i} \frac{\partial u_i}{\partial q^i} + \frac{1}{H_j} \frac{\partial u_j}{\partial q^j} - \frac{u_i}{H_i H_j} \frac{\partial H_i}{\partial q^j} - \right. \\ &\quad \left. - \frac{u_j}{H_i H_j} \frac{\partial H_j}{\partial q^i} + 2\delta_{ij} \frac{u_t}{H_i H_j} \frac{\partial H_j}{\partial q^t} \right). \end{aligned}$$

The Laplacian of a scalar field f is

$$\begin{aligned} \nabla^2 f = \frac{1}{H_1 H_2 H_3} & \left[\frac{\partial}{\partial q^1} \left(\frac{H_2 H_3}{H_1} \frac{\partial f}{\partial q^1} \right) + \right. \\ & + \frac{\partial}{\partial q^2} \left(\frac{H_3 H_1}{H_2} \frac{\partial f}{\partial q^2} \right) + \\ & \left. + \frac{\partial}{\partial q^3} \left(\frac{H_1 H_2}{H_3} \frac{\partial f}{\partial q^3} \right) \right]. \end{aligned} \tag{4.38}$$

In the cylindrical and spherical coordinate systems, for example, (4.36) yields

$$\nabla \cdot \mathbf{f} = \frac{1}{\rho} \frac{\partial}{\partial \rho} (\rho f_\rho) + \frac{1}{\rho} \frac{\partial f_\phi}{\partial \phi} + \frac{\partial f_z}{\partial z}$$

and

$$\nabla \cdot \mathbf{f} = \frac{1}{r^2} \frac{\partial}{\partial r} (r^2 f_r) + \frac{1}{r \sin\theta} \frac{\partial}{\partial \theta} (\sin\theta f_\theta) + \frac{1}{r \sin\theta} \frac{\partial f_\phi}{\partial \phi},$$

while (4.37) yields

$$\nabla \times \mathbf{f} = \hat{\boldsymbol{\rho}} \left(\frac{1}{\rho} \frac{\partial f_z}{\partial \phi} - \frac{\partial f_\phi}{\partial z} \right) + \hat{\boldsymbol{\phi}} \left(\frac{\partial f_\rho}{\partial z} - \frac{\partial f_z}{\partial \rho} \right) + \hat{\mathbf{z}} \frac{1}{\rho} \left(\frac{\partial (\rho f_\phi)}{\partial \rho} - \frac{\partial f_\rho}{\partial \phi} \right)$$

and

$$\begin{aligned} \nabla \times \mathbf{f} = \hat{\mathbf{r}} \frac{1}{r \sin\theta} \left(\frac{\partial}{\partial \theta} (\sin\theta f_\phi) - \frac{\partial f_\theta}{\partial \phi} \right) + \\ + \hat{\boldsymbol{\theta}} \frac{1}{r \sin\theta} \left(\frac{\partial f_r}{\partial \phi} - \sin\theta \frac{\partial (r f_\phi)}{\partial r} \right) + \\ + \hat{\boldsymbol{\phi}} \frac{1}{r} \left(\frac{\partial (r f_\theta)}{\partial r} - \frac{\partial f_r}{\partial \theta} \right). \end{aligned}$$

The specializations of (4.38) to cylindrical and spherical coordinates were given in § 4.8.

4.10 Some Formulas of Integration

Let $f(x_1, x_2, x_3)$ be a continuous function of the Cartesian coordinates x_1, x_2, x_3 in a compact volume V. Let q^1, q^2, q^3 be curvilinear coordinates in the same volume, in one-to-one continuously differentiable correspondence with the Cartesian coordinates, so after transformation of the coordinates we shall write out the same function as $f(q^1, q^2, q^3)$. The transformation is

$$\int_V f(x_1, x_2, x_3)\, dx_1\, dx_2\, dx_3 = \int_V f(q^1, q^2, q^3) J\, dq^1\, dq^2\, dq^3$$

where

$$J = \sqrt{g} = \left| \frac{\partial x_i}{\partial q^j} \right|$$

is the Jacobian.

Exercise 4.21 Show that the Jacobian determinants of the transformations from Cartesian coordinates to cylindrical and spherical coordinates are, respectively,

$$\left| \frac{\partial(x, y, z)}{\partial(\rho, \phi, z)} \right| = \rho, \qquad \left| \frac{\partial(x, y, z)}{\partial(r, \theta, \phi)} \right| = r^2 \sin\theta.$$

Exercise 4.22 Two successive coordinate transformations are given by $x^i = x^i(q^j)$ and $q^i = q^i(\tilde{q}^j)$. Show that the Jacobian determinant of the composite transformation $x^i = x^i(\tilde{q}^j)$ is given by

$$\left|\frac{\partial(x^1, x^2, x^3)}{\partial(\tilde{q}^1, \tilde{q}^2, \tilde{q}^3)}\right| = \left|\frac{\partial(x^1, x^2, x^3)}{\partial(q^1, q^2, q^3)}\right| \left|\frac{\partial(q^1, q^2, q^3)}{\partial(\tilde{q}^1, \tilde{q}^2, \tilde{q}^3)}\right|.$$

For functions $f(x_1, x_2, x_3)$ and $g(x_1, x_2, x_3)$ that are continuously differentiable on a compact volume V with piecewise smooth boundary S, there is the well known Gauss–Ostrogradskij formula for integration by parts:

$$\int_V \frac{\partial f}{\partial x_k} g \, dx_1 \, dx_2 \, dx_3 = -\int_V \frac{\partial g}{\partial x_k} f \, dx_1 \, dx_2 \, dx_3 + \int_S f g \, n_k \, dS,$$

where dS is the differential element of area on S and n_k is the projection of the outward unit normal $\mathbf{n}$ from S onto the axis $\mathbf{i}_k$. In the particular case $g = 1$ we have

$$\int_V \frac{\partial f}{\partial x_k} \, dx_1 \, dx_2 \, dx_3 = \int_S f n_k \, dS. \tag{4.39}$$

Let us use (4.39) to derive some formulas involving the nabla operator that are frequently used in applications. These formulas will be valid in curvilinear coordinates, despite the fact that the intermediate transformations will be done in Cartesian coordinates, because the final results will be written in non-coordinate form. We begin with the integral of ∇f:

$$\int_V \nabla f \, dV = \int_V \mathbf{i}_k \frac{\partial f}{\partial x_k} \, dx_1 \, dx_2 \, dx_3 = \mathbf{i}_k \int_S f n_k \, dS = \int_S f \mathbf{n} \, dS.$$

Now consider an analogous formula for a vector function $\mathbf{f}$:

$$\begin{aligned} \int_V \nabla \mathbf{f} \, dV &= \int_V \mathbf{i}_k \frac{\partial f_t}{\partial x_k} \mathbf{i}_t \, dx_1 \, dx_2 \, dx_3 \\ &= \mathbf{i}_k \mathbf{i}_t \int_S n_k f_t \, dS \\ &= \int_S n_k \mathbf{i}_k f_t \mathbf{i}_t \, dS \\ &= \int_S \mathbf{n} \mathbf{f} \, dS. \end{aligned}$$

Since the left- and right-hand sides are written in non-coordinate form we can use this formula with any coordinate frame with $dV = \sqrt{g} \, dq^1 \, dq^2 \, dq^3$.

This is the formula for a tensor $\nabla \mathbf{f}$; for its trace we have

$$\begin{aligned}\int_V \nabla \cdot \mathbf{f}\, dV &= \int_V \mathbf{i}_k \frac{\partial f_t}{\partial x_k} \cdot \mathbf{i}_t \, dx_1\, dx_2\, dx_3 \\ &= \mathbf{i}_k \cdot \mathbf{i}_t \int_S n_k f_t \, dS \\ &= \int_S n_k \mathbf{i}_k \cdot f_t \mathbf{i}_t \, dS \\ &= \int_S \mathbf{n} \cdot \mathbf{f}\, dS.\end{aligned}$$

In a similar fashion we can derive

$$\int_V \nabla \times \mathbf{f}\, dV = \int_S \mathbf{n} \times \mathbf{f}\, dS.$$

It is easily seen that

$$\int_V \nabla \mathbf{f}^T \, dV = \int_S (\mathbf{nf})^T \, dS = \int_S \mathbf{fn}\, dS.$$

In a similar fashion the reader can use (4.39) to derive the formulas

$$\int_V \nabla \mathbf{A}\, dV = \int_S \mathbf{nA}\, dS,$$

$$\int_V \nabla \cdot \mathbf{A}\, dV = \int_S \mathbf{n} \cdot \mathbf{A}\, dS,$$

and

$$\int_V \nabla \times \mathbf{A}\, dV = \int_S \mathbf{n} \times \mathbf{A}\, dS$$

for a tensor field **A**. Finally let us write out Stokes's formula in non-coordinate form. Recall that this relates a vector function **f** given on a simply-connected surface S to its circulation over the piecewise smooth boundary contour Γ:

$$\oint_\Gamma \mathbf{f} \cdot d\mathbf{r} = \int_S (\mathbf{n} \times \nabla) \cdot \mathbf{f}\, dS.$$

For a tensor **A** of second rank this formula extends to the two formulas

$$\oint_\Gamma d\mathbf{r} \cdot \mathbf{A} = \int_S (\mathbf{n} \times \nabla) \cdot \mathbf{A}\, dS$$

and

$$\oint_\Gamma \mathbf{A} \cdot d\mathbf{r} = \int_S (\mathbf{n} \times \nabla) \cdot \mathbf{A}^T \, dS.$$

We leave these for the reader to prove. We should note that Stokes's formulas hold only when S is simply-connected; for a doubly- or multiply-connected surface, the formulas must be amended by some cyclic constants.

4.11 Norms on Spaces of Vectors and Tensors

Quite frequently we need to characterize the intensity of some vector or tensor field locally or globally. For this the notion of a norm is appropriate. The well known Euclidean norm of a vector $\mathbf{a} = a_k \mathbf{i}_k$ written in a Cartesian frame is

$$\|\mathbf{a}\| = \left(\sum_{k=1}^{3} a_k^2 \right)^{1/2}.$$

This norm relates to the inner product of two vectors $\mathbf{a} = a_k \mathbf{i}_k$ and $\mathbf{b} = b_k \mathbf{i}_k$: we have $\mathbf{a} \cdot \mathbf{b} = a_k b_k$ so that

$$\|\mathbf{a}\| = (\mathbf{a} \cdot \mathbf{a})^{1/2}.$$

In a non-Cartesian frame the components of a vector depend on the lengths of the frame vectors and the angles between them. Since the sum of squared components of a vector depends on the frame we cannot use it as a characteristic of the vector. But the formulas connected with the dot product are invariant under change of frame, so we can use them to characterize the intensity of the vector — its length. Thus for two vectors $\mathbf{x} = x^i \mathbf{e}_i$ and $\mathbf{y} = y^j \mathbf{e}_j$ written in the arbitrary frame, we can introduce a scalar product (i.e., a simple dot product)

$$\mathbf{x} \cdot \mathbf{y} = x^i \mathbf{e}_i \cdot y^j \mathbf{e}_j = x^i y^j g_{ij} = x_i y_j g^{ij} = x^i y_j g_i^j = x^i y_i.$$

Note that only in mixed coordinates does this product resemble the scalar product in a Cartesian frame. Similarly, the norm of a vector $\mathbf{x}$ is

$$\begin{aligned} \|\mathbf{x}\| &= (\mathbf{x} \cdot \mathbf{x})^{1/2} = \left(x^i \mathbf{e}_i \cdot x^j \mathbf{e}_j\right)^{1/2} \\ &= \left(x^i x^j g_{ij}\right)^{1/2} = \left(x_i x_j g^{ij}\right)^{1/2} \\ &= \left(x^i x_i\right)^{1/2}. \end{aligned}$$

This dot product and associated norm have all the properties required from objects of this nature in algebra or functional analysis. Indeed, it is necessary only to check whether all the axioms of the inner product are satisfied. They are

(i) $\mathbf{x} \cdot \mathbf{x} \geq 0$, and $\mathbf{x} \cdot \mathbf{x} = 0$ if and only if $\mathbf{x} = 0$. This property is valid because all the quantities involved can be written in a Cartesian frame where it holds trivially. By the same reasoning we confirm satisfaction of the property

(ii) $\mathbf{x} \cdot \mathbf{y} = \mathbf{y} \cdot \mathbf{x}$. The reader should check that this holds for any representation of the vectors. Finally,

(iii) $(\alpha \mathbf{x} + \beta \mathbf{y}) \cdot \mathbf{z} = \alpha(\mathbf{x} \cdot \mathbf{z}) + \beta(\mathbf{y} \cdot \mathbf{z})$ where α and β are arbitrary real numbers and $\mathbf{z}$ is a vector.

By the general theory then, $\|\mathbf{x}\| = (\mathbf{x} \cdot \mathbf{x})^{1/2}$ satisfies all the axioms of a norm:

(i) $\|\mathbf{x}\| \geq 0$, with $\|\mathbf{x}\| = 0$ if and only if $\mathbf{x} = \mathbf{0}$.
(ii) $\|\alpha \mathbf{x}\| = |\alpha| \, \|\mathbf{x}\|$ for any real α.
(iii) $\|\mathbf{x} + \mathbf{y}\| \leq \|\mathbf{x}\| + \|\mathbf{y}\|$.

In addition we have the Schwarz inequality

$$\|\mathbf{x} \cdot \mathbf{y}\| \leq \|\mathbf{x}\| \, \|\mathbf{y}\| ,$$

where in the case of nonzero vectors the equality holds if and only if $\mathbf{x} = \lambda \mathbf{y}$ for some real λ.

Quite similarly we can introduce a norm and inner product in the set of second rank tensors. We denote the inner product by $(\mathbf{A}, \mathbf{B})$ and define it using the dot product as

$$\begin{aligned}
(\mathbf{A}, \mathbf{B}) &= \mathbf{A} \cdot\cdot\, \mathbf{B}^T \\
&= a^{ij} \mathbf{e}_i \mathbf{e}_j \cdot\cdot\, b^{ts} \mathbf{e}_s \mathbf{e}_t \\
&= a^{ij} b^{ts} g_{js} g_{it} = a_{ij} b_{ts} g^{js} g^{it} \\
&= a^{ij} \mathbf{e}_i \mathbf{e}_j \cdot\cdot\, b_{ts} \mathbf{e}^s \mathbf{e}^t \\
&= a^{ij} b_{ts} g_j^s g_i^t = a^{ij} b_{ij} .
\end{aligned}$$

It is clear that in a Cartesian frame $(\mathbf{A}, \mathbf{A})$ is the sum of all the squared components of $\mathbf{A}$, so this is quite similar to the scalar product of vectors. Using the same reasoning as above we can show that the axioms of the scalar product hold here as well (note that using only a Cartesian frame we could regard the components of a tensor as those of a nine-dimensional vector, so this is another reason why the axioms hold). In this case the inner product axioms are written as

(i) $(\mathbf{A}, \mathbf{A}) \geq 0$, and $(\mathbf{A}, \mathbf{A}) = 0$ if and only if $\mathbf{A} = 0$;
(ii) $(\mathbf{A}, \mathbf{B}) = (\mathbf{B}, \mathbf{A})$;
(iii) $(\alpha\mathbf{X} + \beta\mathbf{Y}, \mathbf{Z}) = \alpha(\mathbf{X}, \mathbf{Z}) + \beta(\mathbf{Y}, \mathbf{Z})$ for any real α, β.

The norm axioms are

(i) $||\mathbf{X}|| \geq 0$, with $||\mathbf{X}|| = 0$ if and only if $\mathbf{X} = \mathbf{0}$;
(ii) $||\alpha\mathbf{X}|| = |\alpha| \, ||\mathbf{X}||$ for any real α;
(iii) $||\mathbf{X} + \mathbf{Y}|| \leq ||\mathbf{X}|| + ||\mathbf{Y}||$.

Hence we still have the Schwarz inequality

$$||(\mathbf{X}, \mathbf{Y})|| \leq ||\mathbf{X}|| \, ||\mathbf{Y}|| \, .$$

In linear algebra many particular implementations of the vector norm can be introduced. We can do the same in any fixed frame, but if we wish to change the frames under consideration we must remember that these norms should change in accordance with the tensor transformation rules.

It is quite easy to extend these notions to tensors of any rank, introducing the inner product as

$$(\mathbf{A}, \mathbf{B}) = a^{i_i i_2 \cdots i_n} b_{i_1 i_2 \cdots i_n} \, .$$

We leave it to the reader to represent this in all of its particular forms and to verify the inner product axioms.

Normed spaces

In textbooks on functional analysis the norm of a vector function is usually introduced in a Cartesian frame. For example, the norm on the space $C(V)$ of continuous vector functions given on a compact V is introduced as

$$||\mathbf{f}(\mathbf{x})||_C = ||f_k(\mathbf{x})\mathbf{i}_k||_C = \max_k \left(\max_V ||f_k(\mathbf{x})|| \right) . \tag{4.40}$$

One cannot use the same formula along with non-Cartesian components since it changes. Moreover, the use of frames having singular points (e.g., the poles of the spherical frame) leads to a situation in which a bounded vector field may have unbounded components. This means that (4.40) is an improper way to characterize the intensity of a vector field. The proper norm for a function given in curvilinear coordinates is based on the above norm:

$$\|\mathbf{f}(\mathbf{x})\| = \left\|f^i(\mathbf{x})\mathbf{r}_i\right\| = \max_V \left(f^i(\mathbf{x})f_i(\mathbf{x})\right)^{1/2}. \tag{4.41}$$

If we would like to use a norm of the type (4.40), we need to remember that during transformation of the frame we must change the form of the norm accordingly.

On the set of second rank tensor functions continuous on a compact V, we can introduce a norm similar to (4.41):

$$\|\mathbf{A}(\mathbf{x})\| = \left\|a^{ij}(\mathbf{x})\mathbf{r}_i\mathbf{r}_j\right\| = \max_V \left(a^{ij}(\mathbf{x})a_{ij}(\mathbf{x})\right)^{1/2}.$$

Finally, note that instead of the norms for continuous vector and tensor functions we can introduce the norms and scalar products corresponding to the space of scalar functions $L^2(V)$. The inner product is then

$$(\mathbf{A}, \mathbf{B}) = \int_V a^{ij} b_{ij}\, dV. \tag{4.42}$$

The reader should verify all the axioms of the inner product for this, and introduce all forms of the inner product and norm corresponding to (4.42).

Exercise 4.23 Let $\mathbf{A}$ be a tensor of the second rank and $\|\mathbf{A}\| = q < 1$. Demonstrate that $\mathbf{E} - \mathbf{A}$ has the inverse $(\mathbf{E} - \mathbf{A})^{-1}$ that is equal to

$$\mathbf{E} + \mathbf{A} + \mathbf{A}^2 + \mathbf{A}^3 + \cdots + \mathbf{A}^n + \cdots.$$

Exercise 4.24 Let $\mathbf{A}$ be a tensor of the second rank and $\|\mathbf{A}\| = q < 1$. What is the inverse to $\mathbf{E} + \mathbf{A}$?

Chapter 5

Elements of Differential Geometry

The standard fare of high school geometry consists mostly of material collected two millennia ago when geometry stood at the center of natural philosophy. The ancient Greeks, however, did not limit their investigations to the circles and straight lines of Euclid's *Elements*. Archimedes, using methods and ideas that were later to underpin the analysis of infinitesimal quantities, could calculate the length of a spiral and the areas and volumes of other complex figures. In elementary algebra we learn to graph simple quadratic functions such as the parabola and hyperbola, and then in analytic geometry we learn to handle space figures such as the ellipsoid. The methods involved are essentially due to Descartes, who connected the ideas of geometry with those of algebra, and their application is largely limited to objects whose describing equations are of the second order. Finally, in elementary calculus we study formulas that permit us to calculate the length of a curve given in Cartesian coordinates, etc. These more powerful methods are now incorporated into a branch of mathematics known as differential geometry.

Differential geometry allows us to characterize curves and figures of a very general nature. The practical importance of this is well illustrated by the problem of optimal pursuit, wherein one object tries to catch another moving object in the shortest possible time.[1] Of course, we shall often make use of standard figures such as circles, parabolas, etc., as specific examples since we are fully familiar with their properties; in this way the objects of both elementary and analytic geometry enter into the more general subject of differential geometry.

[1] A rather humorous statement of one such problem has two old ladies traveling in opposite directions around the base of a hemispherical mountain while a mathematically-minded fly travels back and forth between their noses in the least possible time.

5.1 Elementary Facts from the Theory of Curves

In Chapter 4 we introduced the idea of a coordinate curve in space, which is described by the tip of the radius vector as it moves in such a way that one of the coordinates q^1, q^2, q^3 changes while the other two remain fixed. Now we consider a general curve described in a similar manner by a radius vector whose initial point is the origin and whose terminal point moves through space along the curve. We can describe the position of the radius vector using some parameter t (for a coordinate curve this was the coordinate value q^i). Thus a curve is described as

$$\mathbf{r} = \mathbf{r}(t)$$

where t runs through some set along the real axis. Each value of t corresponds to a point of the curve. Unless otherwise stated we shall suppose that the dependence of $\mathbf{r}$ on t is smooth enough that $\mathbf{r}'(t)$ is continuous in t at each point and, where necessary, that the same holds for $\mathbf{r}''(t)$. Let us mention that for this we need to use the notion of vector norm (§ 4.11). Now we shall denote this norm using usual notation for the magnitude of a vector, e.g., $|\mathbf{r}|$. In the same fashion as we introduced the frame vectors $\mathbf{r}_i$ in Chapter 4 that were tangential to the coordinate lines, we introduce the tangential vector to an arbitrary curve at a point t, which is $\mathbf{r}'(t)$. The differential of the radius vector corresponding to the curve $\mathbf{r} = \mathbf{r}(t)$ is

$$d\mathbf{r}(t) = \mathbf{r}'(t)\, dt.$$

The length of an elementary section of the curve is

$$ds = |\mathbf{r}'(t)|\, dt,$$

and the length of the portion of the curve corresponding to $t \in [a, b]$ is

$$s = \int_a^b |\mathbf{r}'(t)|\, dt.$$

Exercise 5.1 (a) Find the length of one turn of the helix

$$\mathbf{r}(t) = \mathbf{i}_1 \cos t + \mathbf{i}_2 \sin t + \mathbf{i}_3 t.$$

(b) Calculate the perimeter of the ellipse

$$\frac{x^2}{A^2} + \frac{y^2}{B^2} = 1$$

that lies in the $z = 0$ plane.

Exercise 5.2 Show that the general formulas for arc length in the rectangular, cylindrical, and spherical coordinate systems are

$$s = \int_a^b \left[\left(\frac{dx}{dt}\right)^2 + \left(\frac{dy}{dt}\right)^2 + \left(\frac{dz}{dt}\right)^2 \right]^{1/2} dt,$$

$$s = \int_a^b \left[\left(\frac{d\rho}{dt}\right)^2 + \rho^2 \left(\frac{d\phi}{dt}\right)^2 + \left(\frac{dz}{dt}\right)^2 \right]^{1/2} dt,$$

$$s = \int_a^b \left[\left(\frac{dr}{dt}\right)^2 + r^2 \left(\frac{d\theta}{dt}\right)^2 + r^2 \sin^2\theta \left(\frac{d\phi}{dt}\right)^2 \right]^{1/2} dt.$$

Most convenient theoretically is the *natural parameterization* of a curve where the parameter represents the length of the curve calculated from the endpoint:

$$\mathbf{r} = \mathbf{r}(s).$$

With this parameterization

$$ds = |\mathbf{r}'(s)|\, ds \qquad \text{so} \qquad |\mathbf{r}'(s)| = 1.$$

Hence when a curve is parameterized naturally $\boldsymbol{\tau}(s) = \mathbf{r}'(s)$ is the unit tangential vector at point s. In this section s shall denote the length parameter of a curve.

Exercise 5.3 Re-parameterize the helix of Exercise 5.1 in terms of its natural length parameter s. Then calculate the unit tangent to the helix as a function of s.

If there is another parameterization of the curve that relates with the natural parameterization $s = s(t)$, then the vector $d\mathbf{r}(s(t))/dt$ also is tangent to the curve at the point $s = s(t)$. For the derivative of a vector function the chain rule

$$\frac{d\mathbf{r}(s(t))}{dt} = \left.\frac{d\mathbf{r}(s)}{ds}\right|_{s=s(t)} \frac{ds(t)}{dt}$$

holds.

As is known from analytic geometry, when we know the position of a point $\mathbf{a}$ of a straight line and its directional vector $\mathbf{b}$, then the line can be represented in the parametric form

$$\mathbf{r} = \mathbf{a} + \lambda \mathbf{b}.$$

This gives the vectorial equation of the tangent line to a curve at point $\mathbf{r}(t_0)$:

$$\mathbf{r} = \mathbf{r}(t_0) + \lambda \mathbf{r}'(t_0).$$

In Cartesian coordinates (x, y, z) this equation takes the form

$$\begin{aligned} x &= x(t_0) + \lambda x'(t_0), \\ y &= y(t_0) + \lambda y'(t_0), \\ z &= z(t_0) + \lambda z'(t_0), \end{aligned} \tag{5.1}$$

which in nonparametric form is

$$\frac{x - x(t_0)}{x'(t_0)} = \frac{y - y(t_0)}{y'(t_0)} = \frac{z - z(t_0)}{z'(t_0)}. \tag{5.2}$$

Exercise 5.4 Describe the plane curve

$$\mathbf{r}(t) = e^t(\mathbf{i}_1 \cos t + \mathbf{i}_2 \sin t),$$

and find the line tangent to this curve at the point $t = \pi/4$.

Exercise 5.5 (a) Write out the equations for the tangent line to a plane curve corresponding to (5.1) and (5.2) in polar coordinates. (b) Find the angle between the tangent at a point of the curve and the radius vector (from the origin) at this point.

Exercise 5.6 Write out the equations for the tangent line to a curve corresponding to (5.1) and (5.2) in cylindrical and spherical coordinates.

Note that construction of a tangent vector is possible if $\mathbf{r}'(t_0) \neq 0$. A point t_0 where $\mathbf{r}'(t_0) = 0$ is called *singular*.

Exercise 5.7 For a sufficiently smooth curve $\mathbf{r} = \mathbf{r}(t)$ find a tangent at a singular point.

Exercise 5.8 Under the conditions of the previous exercise write out the representation of the type (5.1).

Curvature

For an element of circumference of a circle corresponding to the length Δs, the radius R relates to the central angle $\Delta\varphi$ according to $\Delta s = R\Delta\varphi$. Thus

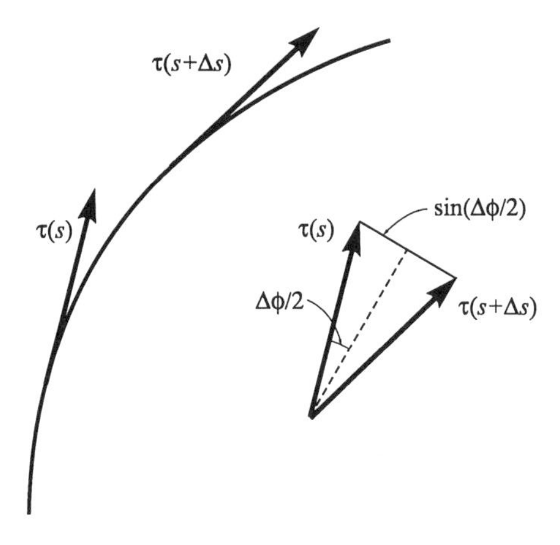

Fig. 5.1 Calculation of curvature.

the *curvature* $k = 1/R$ is

$$k = \frac{\Delta\varphi}{\Delta s}.$$

The curvature of an arbitrary plane curve is defined as the limit

$$k = \lim_{\Delta s \to 0} \frac{\Delta\varphi}{\Delta s}$$

where $\Delta\varphi$ is the change in angle of the tangent to the element of the curve. For a spatial curve the role of the tangent is played by the tangent unit vector $\boldsymbol{\tau}(s)$. When a point moves through the element of the curve corresponding to Δs, the tangent turns through an angle $\Delta\varphi$. Fig. 5.1 demonstrates that $|\boldsymbol{\tau}(s + \Delta s) - \boldsymbol{\tau}(s)|$ is equal to $2\sin(\Delta\varphi/2)$. Since

$$\lim_{\Delta\varphi \to 0} \frac{\sin(\Delta\varphi/2)}{(\Delta\varphi/2)} = 1,$$

the curvature of a spatial curve is given by

$$\begin{aligned} k &= \lim_{\Delta s \to 0} \frac{|\Delta\varphi|}{\Delta s} \\ &= \lim_{\Delta s \to 0} \frac{|\boldsymbol{\tau}(s+\Delta s) - \boldsymbol{\tau}(s)|}{\Delta s} \\ &= \lim_{\Delta s \to 0} \frac{|\mathbf{r}'(s+\Delta s) - \mathbf{r}'(s)|}{\Delta s} \\ &= |\mathbf{r}''(s)|\,. \end{aligned} \tag{5.3}$$

We intentionally introduced the definition of curvature in such a way that it remains nonnegative. For a plane curve the definition is normally given so that k can be positive or negative depending on the sense of convexity ("concave up" or "concave down").

Exercise 5.9 A helix is described by

$$\mathbf{r}(t) = \mathbf{i}_1 \alpha \cos t + \mathbf{i}_2 \alpha \sin t + \mathbf{i}_3 \beta t.$$

Study the curvature as a function of the parameters α and β, showing that $k \to 0$ as $\alpha \to 0$ and $k \to 1/\alpha$ as $\beta \to 0$. Explain.

The moving trihedron

Let us note that $\boldsymbol{\tau}^2 = \boldsymbol{\tau} \cdot \boldsymbol{\tau} = 1$ for all s. This means that

$$\frac{d}{ds}\boldsymbol{\tau}^2(s) = 2\boldsymbol{\tau}(s) \cdot \boldsymbol{\tau}'(s) = 0, \tag{5.4}$$

hence $\boldsymbol{\tau}'(s)$ is orthogonal to $\boldsymbol{\tau}(s)$.

Remark 5.1 It is clear that $\boldsymbol{\tau}(s)$ in (5.4) need not be a tangent vector. Thus we have a general statement: any unit vector $\mathbf{e}(s)$ is orthogonal to its derivative $\mathbf{e}'(s)$. That is, $\mathbf{e}(s) \cdot \mathbf{e}'(s) = 0$.

We define the *principal normal* $\boldsymbol{\nu}$ at point s by the equation

$$\boldsymbol{\nu} = \frac{\mathbf{r}''(s)}{k}. \tag{5.5}$$

If the curve $\mathbf{r} = \mathbf{r}(s)$ lies in a plane, it is clear that $\boldsymbol{\nu}$ lies in the same plane. Any plane through the point s that contains a tangent to the curve at this point is called a tangent plane. Among all the tangent planes at the same point of a non-planar curve, there is a unique one that plays the role of the plane containing a plane curve. It is the plane that contains $\boldsymbol{\tau}$ and $\boldsymbol{\nu}$

simultaneously. This plane is said to be *osculating*, and can be thought of as "the most locally tangent" plane to the curve at the point s. A unit normal to the osculating plane at the same point s is introduced using the relation

$$\boldsymbol{\beta} = \boldsymbol{\tau} \times \boldsymbol{\nu}.$$

These vectors $\boldsymbol{\tau}$, $\boldsymbol{\nu}$, and $\boldsymbol{\beta}$ constitute what is called the *moving trihedron* of the curve. Associated with this frame at each point along the curve is a set of three mutually perpendicular planes. As we stated above, the plane of $\boldsymbol{\tau}$ and $\boldsymbol{\nu}$ is called the osculating plane. The plane of $\boldsymbol{\nu}$ and $\boldsymbol{\beta}$ is called the *normal plane*, and the plane of $\boldsymbol{\beta}$ and $\boldsymbol{\tau}$ is called the *rectifying plane*.

The equation of the osculating plane at point s_0 is

$$(\mathbf{r} - \mathbf{r}(s_0)) \cdot \boldsymbol{\beta}(s_0) = 0$$

where $\mathbf{r}$ is the radius vector of a point of the osculating plane. For general parameterization of the curve $\mathbf{r} = \mathbf{r}(t)$ the equation of the osculating plane in Cartesian coordinates is

$$\begin{vmatrix} x - x(t_0) & y - y(t_0) & z - z(t_0) \\ x'(t_0) & y'(t_0) & z'(t_0) \\ x''(t_0) & y''(t_0) & z''(t_0) \end{vmatrix} = 0.$$

Exercise 5.10 Calculate $\boldsymbol{\nu}$ and $\boldsymbol{\beta}$ for the helix of Exercise 5.9. Then find the equation of the rectifying plane at $t = t_0$.

Using the chain rule we can present the expression for k for an arbitrary parameterization of a curve:

$$k^2 = \frac{(\mathbf{r}'(t) \times \mathbf{r}''(t))^2}{(\mathbf{r}'^2(t))^3}. \tag{5.6}$$

We shall now denote the curvature by k_1, because there is another quantity that characterizes how a curve differs from a straight line. In terms of k_1 we may define the *radius of curvature* as the number $R = 1/k_1$.

Exercise 5.11 Derive expression (5.6).

The principal normal and the binormal to a space curve are not defined uniquely when $\boldsymbol{\tau}'(s) = 0$. Any point at which this condition holds is called a *point of inflection* of the curve. We shall assume that our curves satisfy $\boldsymbol{\tau}'(s) \neq 0$ for all s.

Curves in the plane

The equations of this section take special forms when the curve under consideration lies in a plane. In such a case it is expedient to work in a concrete coordinate system such as rectangular coordinates or plane polar coordinates. In a rectangular coordinate frame $(\hat{\mathbf{x}}, \hat{\mathbf{y}})$ where

$$\mathbf{r} = \hat{\mathbf{x}}x(t) + \hat{\mathbf{y}}y(t),$$

it is easily seen that

$$s = \int_a^b \sqrt{[x'(t)]^2 + [y'(t)]^2}\, dt.$$

The curvature is given by

$$k_1 = \frac{x'(t)y''(t) - x''(t)y'(t)}{\{[x'(t)]^2 + [y'(t)]^2\}^{3/2}}.$$

These formulas correspond to the familiar formulas

$$s = \int_a^b \sqrt{1 + [f'(x)]^2}\, dx, \qquad k_1 = \frac{f''(x)}{\{1 + [f'(x)]^2\}^{3/2}}.$$

from elementary calculus, which are written for a curve expressed in the non-parametric form $y = f(x)$. In polar coordinates where the curve is expressed in the form $r = r(\theta)$, we have

$$s = \int_a^b \sqrt{[r(\theta)]^2 + [r'(\theta)]^2}\, d\theta, \quad k_1 = \frac{[r(\theta)]^2 + 2[r'(\theta)]^2 - r(\theta)r''(\theta)}{\{[r(\theta)]^2 + [r'(\theta)]^2\}^{3/2}}.$$

Note that for a plane curve the curvature k_1 possesses an algebraic sign. We also mention that the general coordinate formulas apply to the case of a plane curve if each index is restricted to take values from the set $\{1, 2\}$.

Exercise 5.12 (a) Find the radius of curvature of the curve $y = x^3$ at the point $(1, 1)$. Repeat for the curve $y = x^4$ at the point $(0, 0)$. (b) Locate the point of maximum curvature of the parabola $y = ax^2 + bx + c$.

Exercise 5.13 Suppose that all the tangents to a smooth curve pass through the same point. Demonstrate the curve is a part of a straight line or the whole line.

Exercise 5.14 Suppose that all the tangents to a smooth curve are parallel to a plane. Show that the curve lies in a plane.

Exercise 5.15 Suppose that all the principal normals of a smooth curve are parallel to a plane. Does this curve lie in a plane? Repeat when all the binormals are parallel to a plane.

5.2 The Torsion of a Curve

When a curve lies in a plane, the binormal $\boldsymbol{\beta}$ is normal to this plane. Moreover, this binormal is normal to the osculating plane to the curve at a point, so the rate of rotation of this plane, which is measured by the rate of turn of the binormal, characterizes how the curve is "non-planar". By analogy to the curvature of a curve we introduce this characteristic of "non-planeness" called the *torsion* or *second curvature*, and define it as the limit of the ratio $\Delta\vartheta/\Delta s$ where $\Delta\vartheta$ is the angle of turn of the binormal $\boldsymbol{\beta}$. Since $\boldsymbol{\beta}$ is a unit vector we can use the same reasoning as we used to derive the expression for the curvature (5.3). Let us denote the torsion by k_2 and write

$$k_2 = \lim_{\Delta s \to 0} \frac{\Delta\vartheta}{\Delta s}.$$

This quantity has an algebraic sign as we explain further below.[2]

Theorem 5.1 *Let* $\mathbf{r} = \mathbf{r}(s)$ *be a three times continuously differentiable vector function of* s*. At any point where* $k_1 \neq 0$*, the torsion of the curve is*

$$k_2 = -\frac{(\mathbf{r}'(s) \times \mathbf{r}''(s)) \cdot \mathbf{r}'''(s)}{k_1^2}.$$

Proof. At a point s where $k_1 \neq 0$ the binormal is defined uniquely as is $\boldsymbol{\nu}$. Since $\boldsymbol{\beta}$ is a unit vector we can define the absolute value of the turn of the binormal, when it moves along the element corresponding to Δs, in the same manner as we used to introduce the curvature of a curve. Now

[2]Basically, a curve having positive torsion will twist in the manner of a right-hand screw thread as s increases.

$|\boldsymbol{\beta}(s+\Delta s)-\boldsymbol{\beta}(s)| = 2\,|\sin \Delta\vartheta/2|$. Thus we have

$$\begin{aligned}
|k_2| &= \lim_{\Delta s\to 0}\left|\frac{\Delta\vartheta}{\Delta s}\right| \\
&= \lim_{\Delta s\to 0}\left|\frac{\Delta\vartheta}{2\sin(\Delta\vartheta/2)}\right|\left|\frac{2\sin(\Delta\vartheta/2)}{\Delta s}\right| \\
&= \lim_{\Delta s\to 0}\left|\frac{\boldsymbol{\beta}(s+\Delta s)-\boldsymbol{\beta}(s)}{\Delta s}\right| \\
&= \left|\boldsymbol{\beta}'(s)\right|.
\end{aligned}$$

Let us demonstrate that $\boldsymbol{\beta}'(s)$ and $\boldsymbol{\nu}$ are parallel. The derivative of any unit vector $\mathbf{x}(s)$ is normal to the vector:

$$0 = \frac{d}{ds}(\mathbf{x}(s)\cdot\mathbf{x}(s)) = 2\mathbf{x}(s)\cdot\mathbf{x}'(s).$$

So $\boldsymbol{\beta}'(s)$ is orthogonal to $\boldsymbol{\beta}(s)$ and is therefore parallel to the osculating plane. Next

$$\boldsymbol{\beta}'(s) = (\boldsymbol{\tau}(s)\times\boldsymbol{\nu}(s))' = \boldsymbol{\tau}'(s)\times\boldsymbol{\nu}(s)+\boldsymbol{\tau}(s)\times\boldsymbol{\nu}'(s) = \boldsymbol{\tau}(s)\times\boldsymbol{\nu}'(s) \quad (5.7)$$

where we used the fact that $\boldsymbol{\tau}'(s)$ is parallel to $\boldsymbol{\nu}(s)$. By (5.7) it follows that $\boldsymbol{\beta}'(s)$ is orthogonal to $\boldsymbol{\tau}(s)$, so we have established the needed property. Since $|\boldsymbol{\nu}(s)| = 1$ it follows that $\left|\boldsymbol{\beta}'(s)\cdot\boldsymbol{\nu}(s)\right| = \left|\boldsymbol{\beta}'(s)\right|$ and thus

$$|k_2| = \left|\boldsymbol{\beta}'(s)\cdot\boldsymbol{\nu}(s)\right|.$$

Let us use the fact that $\boldsymbol{\nu} = \mathbf{r}''/k_1$, and thus $\boldsymbol{\nu}' = \mathbf{r}'''/k_1 - \mathbf{r}''k_1'/k_1^2$ and $\boldsymbol{\beta} = (\mathbf{r}'\times\mathbf{r}'')/k_1$. We get

$$|k_2| = |(\boldsymbol{\tau}\times\boldsymbol{\nu}')\cdot\boldsymbol{\nu}| = \frac{|(\mathbf{r}'\times\mathbf{r}'')\cdot\mathbf{r}'''|}{k_1^2}.$$

Here we used the properties of the mixed product. We now define

$$k_2 = -\frac{(\mathbf{r}'\times\mathbf{r}'')\cdot\mathbf{r}'''}{k_1^2}.$$

The rule for the sign is introduced as follows: if the binormal $\boldsymbol{\beta}$ turns in the direction from $\boldsymbol{\beta}$ to $\boldsymbol{\nu}$, then the sign is positive; otherwise, it is negative. The sign of k_2 is taken in such a way that

$$\boldsymbol{\beta}' = k_2\boldsymbol{\nu}. \quad (5.8)$$

This completes the proof. □

Let us also mention that if we consider another parameterization of the curve, then a simple calculation using the chain rule demonstrates that

$$k_2 = -\frac{(\mathbf{r}'(t) \times \mathbf{r}''(t)) \cdot \mathbf{r}'''(t)}{(\mathbf{r}'(t) \times \mathbf{r}''(t))^2}. \tag{5.9}$$

We have said that the value of the torsion demonstrates the rate at which the curve distinguishes itself from a plane curve. Let us demonstrate this more clearly. Let $k_2 = 0$ for all s. We show that the curve lies in a plane. Indeed $0 = |k_2| = \left|\boldsymbol{\beta}'(s) \cdot \boldsymbol{\nu}(s)\right|$, thus $\left|\boldsymbol{\beta}'(s)\right| = 0$ (since $\boldsymbol{\nu}$ is a unit vector) and so $\boldsymbol{\beta}(s) = \boldsymbol{\beta}^0 = \text{const}$. The tangent vector $\boldsymbol{\tau}$ is orthogonal to $\boldsymbol{\beta}^0$, so $0 = \boldsymbol{\tau} \cdot \boldsymbol{\beta}^0 = \mathbf{r}'(s) \cdot \boldsymbol{\beta}^0$ and thus, integrating, we get $(\mathbf{r}(s) - \mathbf{r}(s_0)) \cdot \boldsymbol{\beta}^0 = 0$. This means that the curve lies in a plane.

By formula (5.8), $k_2 = |d\boldsymbol{\beta}/ds|$. So $k_2 = \lim_{\Delta s \to 0} |\Delta\boldsymbol{\beta}/\Delta s|$. But up to small quantities of the second order of Δs, the increment $\Delta\mathbf{e}$ of a unit vector $\mathbf{e}$ is equal to the angle of rotation of $\mathbf{e}$ when moved through a distance Δs. Thus $\Delta\boldsymbol{\beta}$ is approximately equal to the angle of rotation of the binormal during the shift of the point through Δs, and so k_2 measures the rate of rotation of the binormal when a point moves along the curve.

Exercise 5.16 Demonstrate that a smooth curve lies in a plane only if $k_2 = 0$.

Exercise 5.17 Demonstrate that in Cartesian coordinates

$$k_1 = \frac{\left[(y'z'' - z'y'')^2 + (z'x'' - x'z'')^2 + (x'y'' - y'x'')^2\right]^{1/2}}{\left[(x')^2 + (y')^2 + (z')^2\right]^{3/2}}$$

and

$$k_2 = \frac{\begin{vmatrix} x' & y' & z' \\ x'' & y'' & z'' \\ x''' & y''' & z''' \end{vmatrix}}{(y'z'' - z'y'')^2 + (z'x'' - x'z'')^2 + (x'y'' - y'x'')^2}$$

where the prime denotes d/dt.

Exercise 5.18 What happens to k_1 and k_2 (and the moving trihedron of a curve) if the direction of change of the parameter is reversed ($t \mapsto (-t)$)?

Exercise 5.19 Calculate k_2 for the helix of Exercise 5.9.

5.3 Serret–Frenet Equations

The natural triad $\boldsymbol{\tau}, \boldsymbol{\nu}, \boldsymbol{\beta}$ can serve as coordinate axes of the space; this is used when studying local properties of a curve. Frenet established a system of ordinary differential equations which governs the triad along the curve. We have already derived two of these three equations: they are formulas (5.5) and (5.8). The former will be written as $\boldsymbol{\tau}' = k_1\boldsymbol{\nu}$. Let us derive the third formula of the Serret–Frenet system. We have $\boldsymbol{\nu} = \boldsymbol{\beta} \times \boldsymbol{\tau}$. By this,

$$\begin{aligned}
\boldsymbol{\nu}' &= (\boldsymbol{\beta} \times \boldsymbol{\tau})' \\
&= \boldsymbol{\beta}' \times \boldsymbol{\tau} + \boldsymbol{\beta} \times \boldsymbol{\tau}' \\
&= k_2\boldsymbol{\nu} \times \boldsymbol{\tau} + \boldsymbol{\beta} \times (k_1\boldsymbol{\nu}) \\
&= -k_1\boldsymbol{\tau} - k_2\boldsymbol{\beta}.
\end{aligned}$$

Let us collect the Serret–Frenet equations together:

$$\begin{aligned}
\boldsymbol{\tau}' &= k_1\boldsymbol{\nu}, \\
\boldsymbol{\nu}' &= -k_1\boldsymbol{\tau} - k_2\boldsymbol{\beta}, \\
\boldsymbol{\beta}' &= k_2\boldsymbol{\nu}.
\end{aligned} \tag{5.10}$$

We recall that these equations are written out when the curve has the natural parameterization.

Let us note that if the curvatures $k_1(s)$ and $k_2(s)$ are given functions of s, the system (5.10) becomes a linear system of ordinary differential equations (when written in component form it becomes a system of nine equations in nine unknowns). Fixing some point of the curve in space, by this system we can define $\boldsymbol{\tau}(s)$, $\boldsymbol{\nu}(s)$, and $\boldsymbol{\beta}(s)$ uniquely; then, by the equation $\mathbf{r}'(s) = \boldsymbol{\tau}(s)$, we define the curve $\mathbf{r} = \mathbf{r}(s)$ uniquely as well. Thus $k_1(s)$ and $k_2(s)$ define the curve up to a motion in space. That is why the pair of equations for $k_1(s), k_2(s)$ is called the set of natural equations of the curve.

Let us demonstrate how to use the Serret–Frenet equations to characterize the curve locally. We use the Taylor expansion of the radius vector of a curve at point s:

$$\mathbf{r}(s + \Delta s) = \mathbf{r}(s) + \Delta s\mathbf{r}'(s) + \frac{(\Delta s)^2}{2}\mathbf{r}''(s) + \frac{(\Delta s)^3}{6}\mathbf{r}'''(s) + o(|\Delta s|^3).$$

By the definition $\boldsymbol{\nu} = \mathbf{r}''/k_1$ and by the second of equations (5.10) we get

$$\mathbf{r}''' = (k_1\boldsymbol{\nu})' = k_1'\boldsymbol{\nu} + k_1\boldsymbol{\nu}' = k_1'\boldsymbol{\nu} + k_1(-k_1\boldsymbol{\tau} - k_2\boldsymbol{\beta}).$$

Substituting these we get

$$\mathbf{r}(s+\Delta s) = \mathbf{r}(s)+\Delta s\boldsymbol{\tau}+\frac{(\Delta s)^2}{2}k_1\boldsymbol{\nu}+\frac{(\Delta s)^3}{6}(k_1'\boldsymbol{\nu}-k_1^2\boldsymbol{\tau}-k_1k_2\boldsymbol{\beta})+o(|\Delta s|^3).$$

This shows that, to the order of $(\Delta s)^2$, the curve lies in the osculating plane at point s. At a point s the triad $\boldsymbol{\tau}, \boldsymbol{\nu}, \boldsymbol{\beta}$ is Cartesian. Let us fix this frame and place its origin at $\mathbf{r}(s) = 0$. Defining x, y, z as the components of $\mathbf{r}(s + \Delta s)$ in this frame we obtain the approximate representation for the curve

$$\begin{aligned}
x &= \Delta s - \frac{k_1^2(\Delta s)^3}{6} + o(|\Delta s|^3), \\
y &= \frac{k_1(\Delta s)^2}{2} + \frac{k_1'(\Delta s)^3}{6} + o(|\Delta s|^3), \\
z &= -\frac{k_1k_2(\Delta s)^3}{6} + o(|\Delta s|^3).
\end{aligned}$$

As another application of the Serret–Frenet equations, let us find the velocity and acceleration of a point moving in space. Let s be the length parameter of the trajectory of the point, and let t denote time. We shall use the triad $(\boldsymbol{\tau}, \boldsymbol{\nu}, \boldsymbol{\beta})$ of the trajectory $\mathbf{r} = \mathbf{r}(s)$. The position of the point is given by the equation

$$\mathbf{r} = \mathbf{r}(s(t)).$$

The velocity of the point is

$$\mathbf{v} = \frac{d}{dt}\mathbf{r}(s(t)) = \frac{d\mathbf{r}}{ds}\frac{ds}{dt} = \frac{ds}{dt}\boldsymbol{\tau}.$$

Denoting $v = ds/dt$, the particle speed at each point along its path, we can write $\mathbf{v} = v\boldsymbol{\tau}$. Now let us find the acceleration of the point in the same frame, which is

$$\begin{aligned}
\mathbf{a} &= \frac{d^2}{dt^2}\mathbf{r}(s(t)) = \frac{d}{dt}\mathbf{v}(s(t)) = \frac{d}{dt}\left(\frac{ds}{dt}\boldsymbol{\tau}\right) \\
&= \frac{d^2s}{dt^2}\boldsymbol{\tau} + \frac{ds}{dt}\frac{d\boldsymbol{\tau}}{ds}\frac{ds}{dt} = s''(t)\boldsymbol{\tau} + v^2\frac{d\boldsymbol{\tau}}{ds}.
\end{aligned}$$

Using the Serret–Frenet equations we get

$$\mathbf{a} = s''(t)\boldsymbol{\tau} + k_1v^2\boldsymbol{\nu}$$

where k_1 is the principal curvature of the trajectory. In mechanics this is commonly written as

$$\mathbf{a} = s''(t)\boldsymbol{\tau} + (v^2/\rho)\boldsymbol{\nu}$$

where $\rho = 1/k_1$ is the radius of curvature of the curve. Thus the acceleration vector lies in the osculating plane at each point along the trajectory. On this fact several practical methods of finding the acceleration are based.

Exercise 5.20 Express $d^3\mathbf{r}/ds^3$ in terms of the moving trihedron.

Exercise 5.21 By defining a vector $\boldsymbol{\delta} = k_1\boldsymbol{\beta} - k_2\boldsymbol{\tau}$, show that the Serret–Frenet equations can be written in the form

$$\boldsymbol{\tau}' = \boldsymbol{\delta} \times \boldsymbol{\tau}, \qquad \boldsymbol{\nu}' = \boldsymbol{\delta} \times \boldsymbol{\nu}, \qquad \boldsymbol{\beta}' = \boldsymbol{\delta} \times \boldsymbol{\beta}.$$

The vector $\boldsymbol{\delta}$ is known as the *Darboux vector.*

Exercise 5.22 A particle moves through space in such a way that its position vector is given by

$$\mathbf{r}(t) = \mathbf{i}_1(1 - \cos t) + \mathbf{i}_2 t + \mathbf{i}_3 \sin t.$$

Find the tangential and normal components of the acceleration.

5.4 Elements of the Theory of Surfaces

The reader is familiar with many standard surfaces: the sphere, the cone, the cylinder, etc. These are easy to visualize, and in Cartesian coordinates are described by simple equations. For example, the equation of a sphere reflects the definition of a sphere: all points (x, y, z) of a sphere have the same distance R from the center (x_0, y_0, z_0):

$$(x - x_0)^2 + (y - y_0)^2 + (z - z_0)^2 = R^2. \tag{5.11}$$

An infinite circular cylinder with generator parallel to the z-axis is given by the equation

$$(x - x_0)^2 + (y - y_0)^2 = R^2,$$

in which the variable z does not appear. We can obtain other types of cylindrical surfaces by considering a set of spatial points satisfying the equation

$$f(x, y) = 0.$$

This equation does not depend on z, and thus drawing a generator parallel to the z-axis through the point $(x, y, 0)$ we get a more general cylindrical surface. A paraboloid is an example of a more complex surface:

$$z = ax^2 + 2bxy + cy^2 + dx + ey + f, \tag{5.12}$$

where a, b, c, d, e, f are numerical coefficients. Depending on the values of a, b, c, the paraboloid can be elliptic, hyperbolic or parabolic. The reader may wish to take the time to review these terms from analytic geometry, since we shall use them to characterize the shape of a surface at a point. The areas and volumes associated with all of these standard surfaces (or portions thereof) can be calculated by integration or, sometimes, by the use of elementary formulas.

From a naive viewpoint a surface is something that fully or partially bounds a spatial body. However, a precise definition of the term "surface" is not easy to give. An attempt could be based on a local description, regarding an elementary portion of a surface as a continuous image in space of a small disk in the plane. Unfortunately such a definition would place under consideration surfaces of extremely complex structure. To use the common tools of calculus we need to invoke the idea of smoothness of a surface, even if we do not define this term precisely (it is quite common in the natural sciences to use notions and to study objects that are not explicitly introduced, and about which we know only some things).

As a first step we could view a surface as something in space that resembles the figures mentioned above in the sense that it has zero thickness. To use calculus we must represent the coordinates of the points of a surface using functions. Note that the paraboloid (5.12) can be described using a single function of the two variables x and y, whereas this is not possible with the sphere (5.11). But we can divide a sphere into hemispheres in such a way that each can be described by a separate function of x and y. This can be done with more general surfaces, although it may not be possible to have z as the dependent variable for all portions of a surface. In general, the position of any point of a surface is determined by the values of a pair (u, v) of independent parameters which may or may not be Cartesian coordinates. It is convenient to use the position vector

$$\mathbf{r} = \mathbf{r}(u, v)$$

to locate this point with respect to the coordinate origin. For a Cartesian frame this vector equation is equivalent to the three scalar equations

$$x = x(u, v), \qquad y = y(u, v), \qquad z = z(u, v),$$

but the use of other coordinate frames is common. The pair (u, v) can be regarded as the coordinates of a point in the uv-plane, and a portion of the surface may be regarded as the image of a domain in this plane; in this sense we can say that a surface is two-dimensional. As was the case in earlier sections, we shall consider only sufficiently smooth functions. We have seen that the use of tensor symbolism simplifies many formulas and renders them clear. So instead of the notation (u, v) we shall use (u^1, u^2) to denote the "intrinsic coordinates" of a surface.[3]

One goal of this book is to collect major results and provide explanations of how these can be used by practitioners of many sciences. In particular, we now present formulas used in the mechanics of elastic shells, which relies heavily on the theory of surfaces. First we show how to calculate distances and angles on a general surface. This is done by introducing the "first fundamental form" of the surface. In this way we introduce a metric form of the surface quite analogous to the one used for three-dimensional space. Other surface properties can be described once we possess the notion of the surface normal. The structure of a surface at a point may cause us to experience differences as we depart from the point in various directions. To study this we introduce the "second fundamental form" of the surface. The two fundamental forms provide a full local description of the surface. We shall study various other notions as well, concluding our treatment by adapting tensor analysis to two dimensions and applying the resulting tools to the study of surfaces.

The first fundamental form

We have said that a coordinate line in space is a set of points generated by fixing two of the coordinates q^1, q^2, q^3 and changing the remaining one. In a similar manner we can introduce the concept of a *coordinate level surface*. Such a surface is generated by fixing one of the coordinates q^1, q^2, q^3 and changing the remaining two. A similar idea is used to introduce an arbitrary surface in space. The position of a point on a surface can be characterized by a radius vector initiated from some origin of the space and with terminus that moves along the surface. A point of a surface can be

[3]We use this term to indicate that the coordinates refer to the surface; they are taken in the surface, not from the surrounding space somehow.

characterized by two parameters that we shall denote by u^1, u^2 or sometimes, when convenient, by u, v:

$$\mathbf{r} = \mathbf{r}(u^1, u^2). \tag{5.13}$$

Thus we consider a surface as the spatial image of some domain in the (u^1, u^2) coordinate plane. For simplicity we shall restrict ourselves to those surfaces that are smooth everywhere except possibly for some simple curves or points which lie thereon. So we shall consider the case when $\mathbf{r} = \mathbf{r}(u^1, u^2)$ is as smooth as we like (it is normally necessary for $\mathbf{r} = \mathbf{r}(u^1, u^2)$ to have all of its second (and sometimes third) partial derivatives continuous except for some lines or poles, as is the case with the vertex of a cone). Similar to the case of a three-dimensional space, on a surface there are coordinate lines that arise when in (5.13) we fix one of coordinates. In similar fashion we can introduce the tangent vectors

$$\mathbf{r}_i = \frac{\partial \mathbf{r}(u^1, u^2)}{\partial u^i}, \qquad i = 1, 2.$$

We suppose that, except at some singular lines or poles in the (u^1, u^2) plane, these vectors are continuously differentiable in the coordinates. We also suppose that, except at the same points, these two vectors are not collinear so the vector

$$\mathbf{N} = \mathbf{r}_1 \times \mathbf{r}_2 \neq \mathbf{0}.$$

By this we can introduce the unit normal to the surface

$$\mathbf{n} = \frac{\mathbf{r}_1 \times \mathbf{r}_2}{|\mathbf{r}_1 \times \mathbf{r}_2|},$$

which is simultaneously normal to the plane that is osculating to the surface at point (u^1, u^2). Let us demonstrate this. Let the plane through point (u^1, u^2) have normal $\mathbf{n}$. The osculating plane (Fig. 5.2) is the plane for which

$$\lim_{\Delta s \to 0} \frac{h}{\Delta s} = 0$$

where

$$\Delta s = \left| \mathbf{r}(u^1 + \Delta u^1, u^2 + \Delta u^2) - \mathbf{r}(u^1, u^2) \right|$$

and h is the distance from the point with coordinates $(u^1 + \Delta u^1, u^2 + \Delta u^2)$ to the plane. But

$$h = \left|\left(\mathbf{r}(u^1 + \Delta u^1, u^2 + \Delta u^2) - \mathbf{r}(u^1, u^2)\right) \cdot \mathbf{n}\right|.$$

Using the definition of the differential we have

$$\begin{aligned}\frac{h}{\Delta s} &= \frac{|(\mathbf{r}(u^1 + \Delta u^1, u^2 + \Delta u^2) - \mathbf{r}(u^1, u^2)) \cdot \mathbf{n}|}{|\mathbf{r}(u^1 + \Delta u^1, u^2 + \Delta u^2) - \mathbf{r}(u^1, u^2)|} \\ &= \frac{|(\mathbf{r}_i \Delta u^i + o(|\Delta u^1| + |\Delta u^2|)) \cdot \mathbf{n}|}{|\mathbf{r}_i \Delta u^i + o(|\Delta u^1| + |\Delta u^2|)|} \\ &= \frac{o(|\Delta u^1| + |\Delta u^2|)}{|\mathbf{r}_i \Delta u^i + o(|\Delta u^1| + |\Delta u^2|)|}.\end{aligned}$$

This means that the plane we chose is osculating.

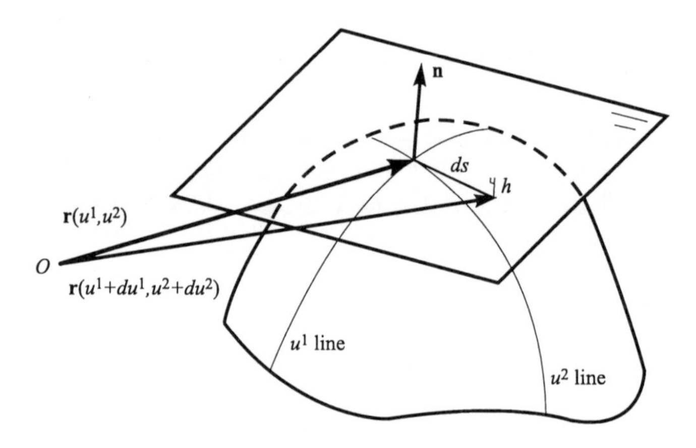

Fig. 5.2 Osculating plane to a surface.

It is clear that much of what was said about $\mathbf{r}_i$ in Chapter 4 should be simply reformulated for the present case. Using the vectors $\mathbf{r}_1, \mathbf{r}_2, \mathbf{n}$, we can find the reciprocal basis in space whose third vector is $\mathbf{n}$ again and whose two others are defined by the equations

$$\mathbf{r}^i \cdot \mathbf{r}_j = \delta^i_j, \quad i = 1, 2, \qquad \mathbf{r}^i \cdot \mathbf{n} = 0.$$

The vectors $\mathbf{r}^i$, $i = 1, 2$, constitute the reciprocal frame to $\mathbf{r}_i$, $i = 1, 2$, in the plane osculating to the surface at (u^1, u^2).

Let us consider the differential

$$d\mathbf{r} = \mathbf{r}_i du^i. \tag{5.14}$$

Here and henceforth, when summing the indices take the values 1 and 2. The differential is the main linear part in du^i of the difference $\Delta\mathbf{r} = \mathbf{r}(u^1 + du^1, u^2 + du^2) - \mathbf{r}(u^1, u^2)$. The main part of the distance between nearby points $\mathbf{r}(u^1 + du^1, u^2 + du^2)$ and $\mathbf{r}(u^1, u^2)$ is given by

$$(ds)^2 = d\mathbf{r} \cdot d\mathbf{r} = \mathbf{r}_i du^i \cdot \mathbf{r}_j du^j = g_{ij} du^i du^j$$

where $g_{ij} = \mathbf{r}_i \cdot \mathbf{r}_j$ is the metric tensor of the surface. The form

$$(ds)^2 = g_{ij} du^i du^j$$

is called the *first fundamental form* of the surface. Beginning with Gauss the metric coefficients were denoted by

$$E = g_{11}, \qquad F = g_{12} = g_{21}, \qquad G = g_{22},$$

so the first fundamental form is

$$(ds)^2 = E(du^1)^2 + 2F du^1 du^2 + G(du^2)^2.$$

Exercise 5.23 Find the first fundamental form of a sphere of radius a.

Besides the length of an elementary curve on the surface, formula (5.14) allows us to find an angle between two elementary curves $d\mathbf{r} = \mathbf{r}_i du^i$ and $d\tilde{\mathbf{r}} = \mathbf{r}_i d\tilde{u}^i$. Indeed, by the definition of the dot product we have

$$\begin{aligned}\cos\varphi &= \frac{d\mathbf{r} \cdot d\tilde{\mathbf{r}}}{|d\mathbf{r}|\,|d\tilde{\mathbf{r}}|} \\ &= \frac{E du^1 d\tilde{u}^1 + F(du^1 d\tilde{u}^2 + du^2 d\tilde{u}^1) + G du^2 d\tilde{u}^2}{\sqrt{E(du^1)^2 + 2F du^1 du^2 + G(du^2)^2}\sqrt{E(d\tilde{u}^1)^2 + 2F d\tilde{u}^1 d\tilde{u}^2 + G(d\tilde{u}^2)^2}}.\end{aligned} \tag{5.15}$$

Exercise 5.24 (a) Find the angle between coordinate lines of a surface. (b) Calculate the angle at which the curve $\theta = \phi$ crosses the equator of a sphere.

The area of the parallelogram that is based on the elementary vectors $\mathbf{r}_1 du^1$ and $\mathbf{r}_2 du^2$ which are tangent to the coordinate lines is

$$dS = \left|\mathbf{r}_1 du^1 \times \mathbf{r}_2 du^2\right|.$$

This approximates up to higher order terms the area of the corresponding curvilinear figure bordered by the coordinate lines

$$u^1 = \text{const}, \quad u^2 = \text{const}, \quad u^1 + du^1 = \text{const}, \quad u^2 + du^2 = \text{const}.$$

Let us find dS in terms of the first fundamental form. We need to calculate $|\mathbf{r}_1 \times \mathbf{r}_2|$. Let us demonstrate that

$$|\mathbf{r}_1 \times \mathbf{r}_2| = \sqrt{EG - F^2}. \tag{5.16}$$

Indeed, by definition of the dot and cross product we have

$$|\mathbf{r}_1 \times \mathbf{r}_2|^2 + (\mathbf{r}_1 \cdot \mathbf{r}_2)^2 = |\mathbf{r}_1|^2 |\mathbf{r}_2|^2 \sin^2 \varphi + |\mathbf{r}_1|^2 |\mathbf{r}_2|^2 \cos^2 \varphi = |\mathbf{r}_1|^2 |\mathbf{r}_2|^2 .$$

Thus

$$|\mathbf{r}_1 \times \mathbf{r}_2|^2 = |\mathbf{r}_1|^2 |\mathbf{r}_2|^2 - (\mathbf{r}_1 \cdot \mathbf{r}_2)^2 = EG - F^2$$

and (5.16) follows.

Summing the elementary areas corresponding to a domain A in the (u^1, u^2) plane and doing the limit passage we get the value

$$S = \int_A \sqrt{EG - F^2}\, du^1\, du^2.$$

It can be shown that for a smooth surface this limit S does not depend on the parameterization and hence is the area of the portion A of the surface.

Exercise 5.25 A cone is described by the position vector

$$\mathbf{r}(u^1, u^2) = \mathbf{i}_1 u^1 \cos u^2 + \mathbf{i}_2 u^1 \sin u^2 + \mathbf{i}_3 u^1,$$

where $0 \le u^1 \le a$ and $0 \le u^2 < 2\pi$. Find the surface area of the cone.

Exercise 5.26 Find the above formulas for a figure of revolution described by the position vector

$$\mathbf{r} = \hat{\mathbf{x}} \rho \cos \phi + \hat{\mathbf{y}} \rho \sin \phi + \hat{\mathbf{z}} f(\rho)$$

where $f(\rho)$ is a suitable profile function.

Exercise 5.27 Let two smooth surfaces be parameterized in such a way that the coefficients of their first fundamental forms are proportional at any point with the same coordinates (so this defines a map of one surface onto the other). Show that the map preserves the angles between corresponding directions on the surfaces.

Geodesics

A fundamental problem in surface theory is to find the curve of minimum length between two given points on a surface S. To treat such a problem it is natural to begin with the first fundamental form

$$(ds)^2 = g_{ij} du^i du^j.$$

We know that parametric equations

$$u^1 = u^1(t), \qquad u^2 = u^2(t),$$

can specify a curve C on S. Writing $(ds)^2$ as

$$(ds)^2 = g_{ij} \frac{du^i}{dt} \frac{du^j}{dt} (dt)^2,$$

the expression for arc length along C from $t = a$ to $t = b$ becomes

$$s = \int_a^b \left(g_{ij} \frac{du^i}{dt} \frac{du^j}{dt} \right)^{1/2} dt. \tag{5.17}$$

We seek a curve of minimum length between a given pair of endpoints: i.e., we seek functions $u^1(t)$ and $u^2(t)$ that minimize s.

Since

$$g_{ij} = g_{ij}(u^1(t), u^2(t)),$$

we see that (5.17) has the form

$$s(\mathbf{u}) = \int_a^b f(t, \mathbf{u}, \dot{\mathbf{u}})\, dt \tag{5.18}$$

where $\mathbf{u} = \mathbf{u}(t) = (u^1(t), u^2(t))$ and the overdot denotes differentiation with respect to t. Because $s(\mathbf{u})$ is a correspondence that assigns a real number s to each function $\mathbf{u}$ in some class of functions, we say that $s(\mathbf{u})$ is a *functional* in $\mathbf{u}$. Here the class of functions includes all admissible routes $\mathbf{u}$ between the given endpoints on the surface.

The minimization of functionals is the subject of a branch of mathematics known as the *calculus of variations.* This subject is rather extensive; fortunately though, the main ideas are simple enough that we can give a very brief treatment oriented toward the present task.

The approach to finding a "minimizer" $\mathbf{u}(t)$ for the functional (5.18) hinges on replacing $\mathbf{u}(t)$ by a new function $\mathbf{u}(t) + \varepsilon\varphi(t)$, where $\varphi(t)$ is an "admissible variation" of the function $\mathbf{u}(t)$ and ε is a small real parameter.

In this discussion we shall consider a variation $\boldsymbol{\varphi}(t)$ to be admissible if $\boldsymbol{\varphi}(a) = \boldsymbol{\varphi}(b) = 0$; that is, if each curve of the form $\mathbf{u}(t) + \varepsilon\boldsymbol{\varphi}(t)$ connects the same endpoints as the curve $\mathbf{u}(t)$. Such a replacement gives

$$s(\mathbf{u} + \varepsilon\boldsymbol{\varphi}) = \int_a^b f(t, \mathbf{u} + \varepsilon\boldsymbol{\varphi}, \dot{\mathbf{u}} + \varepsilon\dot{\boldsymbol{\varphi}})\, dt. \tag{5.19}$$

For fixed $\mathbf{u}$ and $\boldsymbol{\varphi}$ this is a function of the real variable ε, and takes its minimum at $\varepsilon = 0$ for any $\boldsymbol{\varphi}$. Let us vary the components of $\mathbf{u}(t)$ one at a time. If we take $\boldsymbol{\varphi}$ of the special form $\boldsymbol{\varphi}_1(t) = (\varphi(t), 0)$, then (5.19) becomes

$$s(\mathbf{u} + \varepsilon\boldsymbol{\varphi}_1) = \int_a^b f(t, u^1(t) + \varepsilon\varphi(t), u^2(t), \dot{u}^1(t) + \varepsilon\dot{\varphi}(t), \dot{u}^2(t))\, dt.$$

We now set

$$\left. \frac{ds(\mathbf{u} + \varepsilon\boldsymbol{\varphi}_1)}{d\varepsilon} \right|_{\varepsilon=0} = 0 \tag{5.20}$$

and handle the left-hand side using the chain rule:

$$\left. \frac{d}{d\varepsilon} \int_a^b f(t, u^1 + \varepsilon\varphi, u^2, \dot{u}^1 + \varepsilon\dot{\varphi}, \dot{u}^2)\, dt \right|_{\varepsilon=0} =$$
$$= \int_a^b \left[\frac{\partial}{\partial u^1} f(t, u^1, u^2, \dot{u}^1, \dot{u}^2)\varphi + \frac{\partial}{\partial \dot{u}^1} f(t, u^1, u^2, \dot{u}^1, \dot{u}^2)\dot{\varphi} \right] dt.$$

This is known as the *first variation* of the functional $s(\mathbf{u})$ with respect to variations of the form $\mathbf{u}+\varepsilon\boldsymbol{\varphi}_1$; we equate it to zero as in (5.20), and employ integration by parts in the second term of the integrand to get

$$\int_a^b \left[\frac{\partial}{\partial u^1} f(t, u^1, u^2, \dot{u}^1, \dot{u}^2) - \frac{d}{dt}\frac{\partial}{\partial \dot{u}^1} f(t, u^1, u^2, \dot{u}^1, \dot{u}^2) \right] \varphi\, dt = 0.$$

According to the "fundamental lemma" of the calculus of variations, this equation can hold for *all* admissible variations φ only if the bracketed quantity vanishes:

$$\frac{\partial}{\partial u^1} f(t, u^1, u^2, \dot{u}^1, \dot{u}^2) - \frac{d}{dt}\frac{\partial}{\partial \dot{u}^1} f(t, u^1, u^2, \dot{u}^1, \dot{u}^2) = 0.$$

Repeating this process for variations of the special form $\boldsymbol{\varphi}_2(t) = (0, \varphi(t))$ we obtain a similar result

$$\frac{\partial}{\partial u^2} f(t, u^1, u^2, \dot{u}^1, \dot{u}^2) - \frac{d}{dt}\frac{\partial}{\partial \dot{u}^2} f(t, u^1, u^2, \dot{u}^1, \dot{u}^2) = 0.$$

This system of two equations, which we can write in more compact notation as

$$\frac{\partial f}{\partial u^i} - \frac{d}{dt}\frac{\partial f}{\partial \dot{u}^i} = 0, \qquad i = 1, 2, \tag{5.21}$$

is known as the system of *Euler equations* for the functional (5.18). Satisfaction of this system is a necessary condition for $\mathbf{u} = (u^1(t), u^2(t))$ to be a minimizer of s. In the calculus of variations a solution to an Euler equation (or, in this case, a system of such equations) is called an *extremal.* An extremal is analogous to a stationary point of an ordinary function: further testing is required to ascertain whether the extremal actually yields a minimum of the functional (it could yield a maximum, say). We shall refer to solutions of the system (5.21) as *geodesics* on the surface S.

We now use the fact that

$$f^2 = g_{jk}\dot{u}^j\dot{u}^k \tag{5.22}$$

to compute the derivatives in (5.21). First we have

$$2f\frac{\partial f}{\partial u^i} = \frac{\partial g_{jk}}{\partial u^i}\dot{u}^j\dot{u}^k$$

so that

$$\frac{\partial f}{\partial u^i} = \frac{1}{2f}\frac{\partial g_{jk}}{\partial u^i}\dot{u}^j\dot{u}^k. \tag{5.23}$$

Expanding the notation in (5.22) we see that

$$f^2 = g_{11}\dot{u}^1\dot{u}^1 + g_{12}\dot{u}^1\dot{u}^2 + g_{21}\dot{u}^2\dot{u}^1 + g_{22}\dot{u}^2\dot{u}^2,$$

hence

$$2f\frac{\partial f}{\partial \dot{u}^1} = 2g_{11}\dot{u}^1 + g_{12}\dot{u}^2 + g_{21}\dot{u}^2 = 2(g_{11}\dot{u}^1 + g_{12}\dot{u}^2) = 2g_{1j}\dot{u}^j$$

and $\partial/\partial\dot{u}^2$ is taken similarly. So

$$f\frac{\partial f}{\partial \dot{u}^i} = g_{ij}\dot{u}^j,$$

which gives

$$\begin{aligned}\frac{d}{dt}\frac{\partial f}{\partial \dot{u}^i} &= \frac{d}{dt}\left(\frac{g_{ij}\dot{u}^j}{f}\right) \\ &= \frac{1}{f}\frac{d}{dt}(g_{ij}\dot{u}^j) + g_{ij}\dot{u}^j\frac{d}{dt}\left(\frac{1}{f}\right) \\ &= \frac{1}{f}\left(g_{ij}\ddot{u}^j + \dot{u}^j\frac{d}{dt}g_{ij}\right) + g_{ij}\dot{u}^j\left(-\frac{1}{f^2}\frac{df}{dt}\right).\end{aligned}$$

Here

$$\frac{d}{dt}g_{ij} = \frac{\partial g_{ij}}{\partial u^k}\dot{u}^k$$

by the chain rule, so

$$\frac{d}{dt}\frac{\partial f}{\partial \dot{u}^i} = \frac{1}{f}\left(g_{ij}\ddot{u}^j + \frac{\partial g_{ij}}{\partial u^k}\dot{u}^j\dot{u}^k\right) - \frac{g_{ij}\dot{u}^j}{f^2}\frac{df}{dt}.$$

It simplifies things a great deal if we take t to be the arc length parameter; then $f \equiv 1$ and $df/dt \equiv 0$ so that

$$\frac{d}{dt}\frac{\partial f}{\partial \dot{u}^i} = g_{ij}\ddot{u}^j + \frac{\partial g_{ij}}{\partial u^k}\dot{u}^j\dot{u}^k. \tag{5.24}$$

Putting (5.23) and (5.24) into (5.21) we have

$$g_{ij}\ddot{u}^j + \left[\frac{\partial g_{ij}}{\partial u^k} - \frac{1}{2}\frac{\partial g_{jk}}{\partial u^i}\right]\dot{u}^j\dot{u}^k = 0. \tag{5.25}$$

The second term on the left can be rewritten:

$$\begin{aligned}\left[\frac{\partial g_{ij}}{\partial u^k} - \frac{1}{2}\frac{\partial g_{jk}}{\partial u^i}\right] &= \left[\frac{1}{2}\left(\frac{\partial g_{ij}}{\partial u^k} + \frac{\partial g_{ik}}{\partial u^j}\right) - \frac{1}{2}\frac{\partial g_{jk}}{\partial u^i}\right]\dot{u}^j\dot{u}^k \\ &= \frac{1}{2}\left[\frac{\partial g_{ij}}{\partial u^k} + \frac{\partial g_{ik}}{\partial u^j} - \frac{\partial g_{jk}}{\partial u^i}\right]\dot{u}^j\dot{u}^k \\ &= \Gamma_{jki}\dot{u}^j\dot{u}^k.\end{aligned}$$

Putting this into (5.25) and raising the index i, we have finally

$$\ddot{u}^n + \Gamma^n_{jk}\dot{u}^j\dot{u}^k = 0, \qquad n = 1, 2. \tag{5.26}$$

This is a system of two nonlinear differential equations for the unknown $u^n(t)$. Again, t is assumed to be the natural length parameter; this means that the constraint $g_{ij}\dot{u}^j\dot{u}^k = 1$ must be enforced at each value of t along the curve.

As a simple example we may treat a cylinder $\rho = a$. It is clear in advance that if we cut the cylinder along a generator and "unroll" it onto a plane, then the shortest route between two points is the direct segment connecting them. Rolling this plane back into a cylinder we get a curve of the form

$$\phi(t) = c_1 t + c_2, \qquad z(t) = c_3 t + c_4, \tag{5.27}$$

i.e., a helix. Now let us obtain the same result using (5.26). For this particular surface $(u^1, u^2) = (\phi, z)$ and

$$\mathbf{r} = \hat{\mathbf{x}} a \cos\phi + \hat{\mathbf{y}} a \sin\phi + \hat{\mathbf{z}} z.$$

We find

$$\frac{\partial \mathbf{r}}{\partial \phi} = -\hat{\mathbf{x}} a \sin\phi + \hat{\mathbf{y}} a \cos\phi, \qquad \frac{\partial \mathbf{r}}{\partial z} = \hat{\mathbf{z}},$$

giving $g_{11} = a^2$, $g_{12} = g_{21} = 0$, $g_{22} = 1$. Since these are all constants the Christoffel coefficients are all zero, and (5.26) reduces to the system

$$\ddot{\phi} = 0, \qquad \ddot{z} = 0.$$

Integration replicates the result (5.27).

Exercise 5.28 Find the geodesics of a sphere. Show that they are great circles of the sphere.

This exercise shows that a geodesic is not always a shortest route, since we can get either part of a great circle on the sphere connecting two given points, and both of them are geodesics.

5.5 The Second Fundamental Form of a Surface

Using the Taylor expansion we can find the change of $\mathbf{r}(u^1, u^2)$ to a higher degree of approximation:

$$\mathbf{r}(u^1 + \Delta u^1, u^2 + \Delta u^2) = \mathbf{r}(u^1, u^2) + \frac{\partial \mathbf{r}(u^1, u^2)}{\partial u^i} \Delta u^i + \\ + \frac{1}{2} \frac{\partial^2 \mathbf{r}(u^1, u^2)}{\partial u^i \partial u^j} \Delta u^i \Delta u^j + o\left((\Delta u^1)^2 + (\Delta u^2)^2\right).$$

Here we apply the o-notation to vector quantities, indicating that the norm of the remainder is of higher order of smallness. We are interested in the deviation of the surface from the osculating plane in a neighborhood of a point (u^1, u^2). This value is given through the dot product by the normal $\mathbf{n}$:

$$\begin{aligned}
&(\mathbf{r}(u^1 + \Delta u^1, u^2 + \Delta u^2) - \mathbf{r}(u^1, u^2)) \cdot \mathbf{n} \\
&= \frac{\partial \mathbf{r}(u^1, u^2)}{\partial u^i} \cdot \mathbf{n} \Delta u^i + \frac{1}{2} \frac{\partial^2 \mathbf{r}(u^1, u^2)}{\partial u^i \partial u^j} \cdot \mathbf{n} \Delta u^i \Delta u^j + o\left((\Delta u^1)^2 + (\Delta u^2)^2\right) \\
&= \frac{1}{2} \frac{\partial^2 \mathbf{r}(u^1, u^2)}{\partial u^i \partial u^j} \cdot \mathbf{n} \Delta u^i \Delta u^j + o\left((\Delta u^1)^2 + (\Delta u^2)^2\right).
\end{aligned}$$

We have used the fact that $\mathbf{r}_i \cdot \mathbf{n} = 0$. The terms of the second order of smallness in Δu^i constitute what is called the *second fundamental form.* Denoting

$$\begin{aligned}
\frac{\partial^2 \mathbf{r}(u^1, u^2)}{(\partial u^1)^2} \cdot \mathbf{n} &= L(u^1, u^2) = L, \\
\frac{\partial^2 \mathbf{r}(u^1, u^2)}{\partial u^1 \partial u^2} \cdot \mathbf{n} &= M(u^1, u^2) = M, \\
\frac{\partial^2 \mathbf{r}(u^1, u^2)}{(\partial u^2)^2} \cdot \mathbf{n} &= N(u^1, u^2) = N,
\end{aligned}$$

we introduce the second fundamental form by

$$d^2\mathbf{r} \cdot \mathbf{n} = L(du^1)^2 + 2M du^1 du^2 + N(du^2)^2. \tag{5.28}$$

The deviation of the surface from the osculating plane in a small neighborhood of (u^1, u^2) is given by the following approximation:

$$z = \frac{1}{2}\left(L(du^1)^2 + 2M du^1 du^2 + N(du^2)^2\right), \tag{5.29}$$

and so (5.29) defines the nature of the behavior of the surface in this neighborhood. This formula defines a paraboloid

$$z = \frac{1}{2}\left(L(v^1)^2 + 2M v^1 v^2 + N(v^2)^2\right) \tag{5.30}$$

written in the triad $\left(\mathbf{r}_1(u^1, u^2), \mathbf{r}_2(u^1, u^2), \mathbf{n}(u^1, u^2)\right)$ when the origin is the point of the surface with coordinates (u^1, u^2). Depending on the coefficients L, M, N, this paraboloid can be elliptic, hyperbolic or parabolic. If $L = M = N = 0$ then the corresponding point is called a planar point.

For the second fundamental form of the surface there is a representation different from (5.28). Differentiating the identity $\mathbf{r}_i \cdot \mathbf{n} = 0$ we get

$$\frac{\partial \mathbf{r}_i}{\partial u^j} \cdot \mathbf{n} = -\mathbf{r}_i \cdot \frac{\partial \mathbf{n}}{\partial u^j}. \tag{5.31}$$

Dot multiplying $d\mathbf{r} \cdot d\mathbf{n}$ we get

$$\mathbf{r}_i du^i \cdot \frac{\partial \mathbf{n}}{\partial u^j} du^j = -\left(\frac{\partial \mathbf{r}_i}{\partial u^j} du^i du^j\right) \cdot \mathbf{n} = -d^2\mathbf{r} \cdot \mathbf{n},$$

which means that

$$-d\mathbf{r} \cdot d\mathbf{n} = L(du^1)^2 + 2M du^1 du^2 + N(du^2)^2.$$

Exercise 5.29 Let a surface be given in Cartesian components as

$$\mathbf{r} = x(u,v)\hat{\mathbf{x}} + y(u,v)\hat{\mathbf{y}} + z(u,v)\hat{\mathbf{z}}.$$

Let the subscripts u and v denote corresponding partial derivatives with respect to u and v. Demonstrate that the coefficients of the first fundamental form are

$$\begin{aligned} E &= \mathbf{r}_u^2 = x_u^2 + y_u^2 + z_u^2, \\ F &= \mathbf{r}_u \cdot \mathbf{r}_v = x_u x_v + y_u y_v + z_u z_v, \\ G &= \mathbf{r}_v^2 = x_v^2 + y_v^2 + z_v^2. \end{aligned}$$

Then show that the coefficients of the second fundamental form are

$$L = \frac{\begin{vmatrix} x_{uu} & y_{uu} & z_{uu} \\ x_u & y_u & z_u \\ x_v & y_v & z_v \end{vmatrix}}{\sqrt{EG - F^2}}, \quad M = \frac{\begin{vmatrix} x_{uv} & y_{uv} & z_{uv} \\ x_u & y_u & z_u \\ x_v & y_v & z_v \end{vmatrix}}{\sqrt{EG - F^2}}, \quad N = \frac{\begin{vmatrix} x_{vv} & y_{vv} & z_{vv} \\ x_u & y_u & z_u \\ x_v & y_v & z_v \end{vmatrix}}{\sqrt{EG - F^2}}.$$

Exercise 5.30 Find L, M, N for a surface of revolution.

The normal curvature of the surface

Let us consider a curve lying on the surface. The curve can be uniquely defined by a parameter s in such a way that $u^1 = u^1(s)$, $u^2 = u^2(s)$; for simplicity we assume s to be the length parameter. Then the equation of the curve is $\mathbf{r} = \mathbf{r}(u^1(s), u^2(s)) = \mathbf{r}(s)$, and a tangential vector to the curve at (u^1, u^2) is $(\mathbf{r}_i du^i/ds)ds$. Consider the curvature of this curve, which can be found using $\mathbf{r}''(s)$: the normal $\boldsymbol{\nu}$ to the curve has the same direction and k_1 equals $|\mathbf{r}''(s)|$. It follows that

$$\mathbf{r}''(s) \cdot \mathbf{n} = k_1 \cos\vartheta \tag{5.32}$$

where ϑ is the angle between $\boldsymbol{\nu}$ and $\mathbf{n}$. Let us rewrite (5.32) in the form

$$k_1 \cos\vartheta =$$

$$= \frac{\frac{\partial^2\mathbf{r}}{(\partial u^1)^2}\left(\frac{du^1}{ds}\right)^2 (ds)^2 + 2\frac{\partial^2\mathbf{r}}{\partial u^1\partial u^2}\frac{du^1}{ds}\frac{du^2}{ds}(ds)^2 + \frac{\partial^2\mathbf{r}}{(\partial u^2)^2}\left(\frac{du^2}{ds}\right)^2 (ds)^2}{(ds)^2}\cdot\mathbf{n}$$

$$= \frac{L(du^1(s))^2 + 2M du^1(s)du^2(s) + N(du^2(s))^2}{E(du^1(s))^2 + 2F du^1(s)du^2(s) + G(du^2(s))^2}. \tag{5.33}$$

For all the curves with the same direction defined by $du^1(s), du^2(s)$ the right-hand side of (5.33) is the same as the ratio of the second to the first fundamental forms. Define this value as k_0 and call it the *normal curvature* of the surface in the direction $du^1(s), du^2(s)$. Geometrically, this is the curvature of the curve formed by intersecting the surface with the plane through $(u^1(s), u^2(s))$ that is parallel to $\mathbf{n}$ and $\mathbf{r}_i du^i(s)$. This normal curvature satisfies the equality

$$k_0 = k_1 \cos\vartheta$$

which is the main part of *Meusnier's theorem*. From this formula for curvatures it follows that the normal curvature of the surface in the direction $du^1 : du^2$ has minimal absolute value among all the curvatures of the curves on the surface through $(u^1(s), u^2(s))$ having the same direction $du^1 : du^2$. Let us note that the curvature of the osculating paraboloid is characterized by the same equation. The normal curvature depends on the direction $du^1 : du^2 = x : y$ only. Let us rewrite its expression as

$$k_0 = \frac{Lx^2 + 2Mxy + Ny^2}{Ex^2 + 2Fxy + Gy^2}. \tag{5.34}$$

The right-hand side of (5.34) is a homogeneous form with respect to x, y of zero order; it is easy to see that the minimum and maximum of k_0 can be found as the minimum and maximum of the quadratic form

$$Lx^2 + 2Mxy + Ny^2$$

when

$$Ex^2 + 2Fxy + Gy^2 = 1. \tag{5.35}$$

This problem can be solved with use of the Lagrange multiplier. On the curve (5.35) denote

$$\min(Lx^2 + 2Mxy + Ny^2) = k_{\min}, \quad \max(Lx^2 + 2Mxy + Ny^2) = k_{\max}.$$

Let the corresponding values for x, y be (x_1, y_1) and (x_2, y_2), respectively. If $k_{\min} \neq k_{\max}$ these values define only two directions and it can be shown that

$$Ex_1x_2 + F(x_1y_2 + x_2y_1) + Gy_1y_2 = 0,$$

which, by (5.15), means that the directions corresponding to the extreme values of the normal curvature are mutually orthogonal. In the case when $k_{\min} = k_{\max}$ the normal curvature is the same in all directions (the osculating paraboloid corresponds to a paraboloid of revolution). These $k_{\min}, k_{\max}$ which are the extreme normal curvatures of the surface at a point are the characteristics of the surface at a point and so they are invariant with respect to the change of coordinates of the surface. So are two other characteristics of the surface at a point:

$$H = \frac{1}{2}(k_{\min} + k_{\max}),$$

called the *mean curvature* of the surface, and

$$K = k_{\min}k_{\max},$$

called the *Gaussian* or *complete curvature* of the surface. We present without proof the equations in terms of the fundamental forms of the surface; they are

$$H = \frac{1}{2}\frac{LG - 2MF + NE}{EG - F^2}, \qquad K = \frac{LN - M^2}{EG - F^2}.$$

Since $EG - F^2 > 0$, the sign of K is the sign of $LN - M^2$. When $K = 0$ everywhere the surface is developable.[4] These and many other facts on the properties of surfaces with given H and K can be found in any textbook on differential geometry.

Exercise 5.31 Find the mean and Gaussian curvatures for a sphere of radius a.

Exercise 5.32 Find the mean and Gaussian curvatures at $x = y = 0$ of the following paraboloids: (a) $z = axy$; (b) $z = a(x^2+y^2)$; (c) $z = ax^2+by^2$.

[4]Roughly speaking, a *developable surface* is a surface that can be flattened into a portion of a plane without compressing or stretching any part of it. Examples include cones and cylinders, but not spheres. A developable surface can be generated by sweeping a straight line (*generator*) along a curve through space. We will give a more precise definition on page 137.

Exercise 5.33 Suppose a surface is given in Cartesian coordinates:

$$\mathbf{r} = \hat{\mathbf{x}}x + \hat{\mathbf{y}}y + \hat{\mathbf{z}}f(x,y).$$

Demonstrate that (a)

$$E = \mathbf{r}_x^2 = 1 + f_x^2, \qquad F = \mathbf{r}_x \cdot \mathbf{r}_y = f_x f_y, \qquad G = \mathbf{r}_y^2 = 1 + f_y^2;$$

(b)

$$EG - F^2 = 1 + f_x^2 + f_y^2;$$

(c)

$$L = \frac{f_{xx}}{\sqrt{1 + f_x^2 + f_y^2}}, \qquad M = \frac{f_{xy}}{\sqrt{1 + f_x^2 + f_y^2}}, \qquad N = \frac{f_{yy}}{\sqrt{1 + f_x^2 + f_y^2}};$$

(d) the area of a portion of the surface is

$$S = \int_D \sqrt{1 + f_x^2 + f_y^2}\, dx\, dy;$$

(e) the Gaussian curvature is

$$K = \frac{LN - M^2}{EG - F^2} = \frac{f_{xx}f_{yy} - f_{xy}^2}{\left(1 + f_x^2 + f_y^2\right)^2}.$$

5.6 Derivation Formulas

At each point of a surface there is defined the frame triad

$$\mathbf{r}_1(u^1, u^2),\ \mathbf{r}_2(u^1, u^2),\ \mathbf{n}(u^1, u^2).$$

In the theory of shells they introduce the curvilinear coordinates in a neighborhood of the surface in such a way that on the surface the triad $(\mathbf{r}_1, \mathbf{r}_2, \mathbf{n})$ is preserved. Since we need to find different characteristics of fields given on the surface and outside of it we need to find the derivatives of the triad with respect to the coordinates. The goal of this section is to present these in terms of the surface we have introduced: that is, in terms of the first and second fundamental forms of the surface.

We start with the representation for the derivatives through the Christoffel notation. This is valid because, by assumption, $(\mathbf{r}_1, \mathbf{r}_2, \mathbf{n})$ is a basis of the space of vectors. So

$$\mathbf{r}_{ij} = \frac{\partial^2 \mathbf{r}}{\partial u^i \partial u^j} = \Gamma^t_{ij}\mathbf{r}_t + \lambda_{ij}\mathbf{n}. \tag{5.36}$$

We recall that we use indices i, j, t taking values from the set $\{1, 2\}$, which is why we introduce the notation λ_{ij} for the coefficients at $\mathbf{n}$. Similarly let us introduce the expansion for the derivatives of $\mathbf{n}$. To derive the coefficients λ_{ij} we dot multiply (5.36) by $\mathbf{n}$. Using the expressions for the second fundamental form we get

$$\lambda_{11} = L, \qquad \lambda_{12} = \lambda_{21} = M, \qquad \lambda_{22} = N.$$

Next, dot multiplying (5.36) first by $\mathbf{r}_1$ and then by $\mathbf{r}_2$ we get six equations in the six unknown Christoffel symbols:

$$\begin{aligned}
\frac{1}{2}\frac{\partial E}{\partial u^1} &= \Gamma^1_{11}E + \Gamma^2_{11}F,\\
\frac{\partial F}{\partial u^1} - \frac{1}{2}\frac{\partial E}{\partial u^2} &= \Gamma^1_{11}F + \Gamma^2_{11}G,\\
\frac{1}{2}\frac{\partial E}{\partial u^2} &= \Gamma^1_{12}E + \Gamma^2_{12}F,\\
\frac{1}{2}\frac{\partial G}{\partial u^1} &= \Gamma^1_{12}F + \Gamma^2_{12}G,\\
\frac{\partial F}{\partial u^2} - \frac{1}{2}\frac{\partial G}{\partial u^1} &= \Gamma^1_{22}E + \Gamma^2_{22}F,\\
\frac{1}{2}\frac{\partial G}{\partial u^2} &= \Gamma^1_{22}F + \Gamma^2_{22}G.
\end{aligned}$$

This system can be easily solved for the Christoffel coefficients since the system splits into pairs of equations with respect to the pairs of Christoffel symbols. The first pair yields, for instance,

$$\Gamma^1_{11} = \frac{1}{EG - F^2}\left[\frac{G}{2}\frac{\partial E}{\partial u^1} - F\frac{\partial F}{\partial u^1} + \frac{F}{2}\frac{\partial E}{\partial u^2}\right],$$

and a similar expression for Γ^2_{11}. We leave the rest of this work to the reader, mentioning that the expressions for the Christoffel symbols are composed only of the coefficients of the first fundamental form of the surface. Now let us consider the expressions for the derivatives of the normal $\mathbf{n}$:

$$\mathbf{n}_i = \frac{\partial \mathbf{n}}{\partial u^i} = \mu_i^{\cdot t}\mathbf{r}_t + \mu_i^{\cdot 3}\mathbf{n}. \tag{5.37}$$

Let us find the coefficients of these expansions through the coefficients of the fundamental forms of the surface. Because of the equality

$$0 = \frac{\partial}{\partial u^i} 1 = \frac{\partial}{\partial u^i}(\mathbf{n}\cdot\mathbf{n}) = 2\mathbf{n}_i \cdot \mathbf{n}$$

we have

$$\mu_i^{\cdot 3} = 0.$$

Let us dot multiply (5.37) when $i = 1$ by $\mathbf{r}_1$ and $\mathbf{r}_2$ successively. Using the definitions for the coefficients of the first and second fundamental forms we have the equations

$$-L = \mu_1^{\cdot 1} E + \mu_1^{\cdot 2} F, \qquad -M = \mu_1^{\cdot 1} F + \mu_1^{\cdot 2} G,$$

from which we get

$$\mu_1^{\cdot 1} = \frac{-LG + MF}{EG - F^2}, \qquad \mu_1^{\cdot 2} = \frac{LF - ME}{EG - F^2}.$$

Repeating this procedure for (5.37) when $i = 2$ we similarly obtain

$$\mu_2^{\cdot 1} = \frac{NF - MG}{EG - F^2}, \qquad \mu_2^{\cdot 2} = \frac{-NE + MF}{EG - F^2}.$$

Exercise 5.34 Derive the Christoffel symbols for the case when the first fundamental form is $(du^1)^2 + G(du^2)^2$.

Exercise 5.35 For a surface of revolution with coordinate lines being parallels and meridians, derive all coefficients of the expansions (5.36) and (5.37).

Some useful formulas

There are useful identities that are frequently employed. Their derivation is lengthy so we do not present it here. Let us redenote the coefficients of the second fundamental form using index notations:

$$b_{11} = L, \qquad b_{12} = b_{21} = M, \qquad b_{22} = N.$$

These are the covariant components of a corresponding symmetric tensor of second rank. The first formula is due to Gauss:

$$b_{11}b_{22} - b_{12}^2 = \frac{\partial^2 g_{12}}{\partial u^1 \partial u^2} - \frac{1}{2}\left(\frac{\partial^2 g_{11}}{(\partial u^2)^2} + \frac{\partial^2 g_{22}}{(\partial u^1)^2}\right) + \left(\Gamma_{12}^i \Gamma_{12}^j - \Gamma_{11}^i \Gamma_{22}^j\right) g_{ij},$$

from which (and the form of K, see page 121) it follows that the Gaussian curvature of the surface can be expressed purely in terms of the first fundamental form. The next formulas are due to Peterson and Codazzi:

$$\frac{\partial b_{i1}}{\partial u^2} - \frac{\partial b_{i2}}{\partial u^1} = \Gamma^t_{i2} b_{t1} - \Gamma^t_{i1} b_{t2}, \qquad i = 1, 2.$$

Using the formulas of this paragraph we can derive the formulas for differentiating a spatial vector $\mathbf{u} = u^i \mathbf{r}_i + u_3 \mathbf{n} = u_i \mathbf{r}^i + u_3 \mathbf{n}$ given on the surface. Denoting $\mathbf{r}_3 = \mathbf{r}^3 = \mathbf{n}$, we can derive the following formulas that mimic the formulas of Chapter 4:

$$\frac{\partial \mathbf{u}}{\partial u^i} = \nabla_i u^t \mathbf{r}_t = \nabla_i u_t \mathbf{r}^t.$$

In these, the covariant derivatives are

$$\begin{aligned}
\nabla_i u^j &= \frac{\partial u^j}{\partial u^i} + \Gamma^j_{it} u^t - B^j_i u_3, \\
\nabla_i u_j &= \frac{\partial u_j}{\partial u^i} + \Gamma^t_{ij} u_t - B_{ij} u_3, \\
\nabla_i u_3 &= \frac{\partial u_3}{\partial u^i} + B_{it} u^t = \frac{\partial u_3}{\partial u^i} + B^t_i u_t.
\end{aligned}$$

5.7 Implicit Representation of a Curve; Contact of Curves

Our work in Chapter 5 has centered around technical topics needed to derive equations describing certain natural objects. We now turn to some less technical but quite important questions that can further our understanding of differential geometry.

We have described a curve by the representation $\mathbf{r} = \mathbf{r}(t)$. From this point of view, a curve is the image of some one-dimensional domain of the parameter t. Similarly, a surface was an image of a two-dimensional domain, given by a formula

$$\mathbf{r} = \mathbf{r}(u^1, u^2). \tag{5.38}$$

However, we know that in Cartesian coordinates a sphere is described by the equation

$$(x - x_0)^2 + (y - y_0)^2 + (z - z_0)^2 = R^2,$$

and this cannot be represented in the form (5.38) for all points simultaneously. This is an example of the implicit form of description of a surface.

We can describe any small part of a sphere in an explicit form of the type (5.38). This is a point from which the theory of manifolds arose: a surface such as a sphere can be divided into small portions, each of which can be described in the needed way.

However, implicit form descriptions of surfaces and curves are quite common in practice. Let us suppose that the equation of a surface is

$$F(\mathbf{r}) = 0.$$

In Cartesian coordinates this would look like

$$F(x, y, z) = 0.$$

Earlier we described coordinate curves in space as sets given by the equation $\mathbf{r} = \mathbf{r}(\alpha^1, \alpha^2, \alpha^3)$ when two of the three coordinate parameters $\alpha^1, \alpha^2, \alpha^3$ are fixed. Each such curve can be described alternatively as the intersection of two coordinate surfaces. Consider for instance an α^1 coordinate curve — the curve corresponding to $\alpha^2 = \alpha_0^2$, $\alpha^3 = \alpha_0^3$, is the intersection of the surfaces $\mathbf{r} = \mathbf{r}(\alpha^1, \alpha^2, \alpha_0^3)$ and $\mathbf{r} = \mathbf{r}(\alpha^1, \alpha_0^2, \alpha^3)$. Similarly a curve in space can be represented as the set of points described by a vector $\mathbf{r}$ that satisfies two simultaneous equations

$$F_1(\mathbf{r}) = 0, \qquad F_2(\mathbf{r}) = 0,$$

or, in coordinate form,

$$F_1(\alpha^1, \alpha^2, \alpha^3) = 0, \qquad F_2(\alpha^1, \alpha^2, \alpha^3) = 0.$$

A tangent $\mathbf{t}$ to the curve at a point must be orthogonal to the two surface normals at the point; if $\mathbf{n}_1$ and $\mathbf{n}_2$ are these normals, we can write

$$\mathbf{t} = \mathbf{n}_1 \times \mathbf{n}_2.$$

Contact of curves

A curve in the plane can be also shown implicitly, but using just one equation:

$$F(\mathbf{r}) = 0 \qquad \text{or} \qquad F(\alpha^1, \alpha^2) = 0.$$

We would like to study the problem of approximating a given plane curve, at a given point, by another curve taken from a parametric family of curves. Let us begin with the following problem:

> Given a smooth curve A and point C on it, select from all straight lines in the plane that which best approximates the behavior of A at C.

Of course, one solution is the line tangent to A at C. We see this by considering a line through C that intersects A at a point D close to C, and then producing a limit passage under which this line rotates about C in such a way that D tends to C. The line approached in the limit is the tangent line.

Let us extend this idea somewhat and seek an approximation that can reflect both the direction of the curve A at C and its curvature at that point. Since straight lines cannot reflect curvature behavior, we shall have to employ some other family of approximating curves. We could use the family of all circles in the plane, since any three points in the plane determine a unique circle. Taking two points D_1 and D_2 near C and drawing the circle through these three points, we could attempt a limit passage under which both D_1 and D_2 tend to C. This should yield a circle capable of reflecting both desired properties (tangent and curvature) of A at C. The family of all circles in the plane is described in Cartesian coordinates by the equation

$$(x - x_0)^2 + (y - y_0)^2 = R^2, \tag{5.39}$$

where the three values x_0, y_0 and R are free parameters that should be properly chosen to approximate a given curve at a point. So we see that a three-parameter family of plane curves will be needed to get the better approximation we seek. We could employ a parabola from the family $y = ax^2 + bx + c$, but this manner of approximation would be less informative.

In general we can seek to approximate a plane curve

$$\mathbf{r} = \mathbf{r}(t) \tag{5.40}$$

at a point $t = t_0$ by introducing a parametric family Φ_n of curves

$$F(\mathbf{r}, a_1, \ldots, a_n) = 0,$$

where the $a_1, \ldots, a_n$ are free parameters.[5] Suppose we take n points

$$\mathbf{r}(t_k) \qquad k = 0, \ldots, n-1$$

[5] We suppose that the curve (5.40) is sufficiently smooth at t_0, as is each curve taken from the family Φ_n.

of the curve (5.40), where all the t_k are close to t_0. With n free parameters at our disposal we should be able to find a curve from Φ_n that goes through these points. This means that the system

$$\begin{aligned} F(\mathbf{r}(t_0), a_1, \ldots, a_n) &= 0, \\ F(\mathbf{r}(t_1), a_1, \ldots, a_n) &= 0, \\ &\vdots \\ F(\mathbf{r}(t_{n-1}), a_1, \ldots, a_n) &= 0, \end{aligned} \tag{5.41}$$

is satisfied by some set of parameters $a_1, \ldots, a_n$. We assume that the a_i depend on the t_k in such a way that when all the t_k tend to t_0 the a_i tend continuously to some respective values b_i. Thus we find the curve $F(\mathbf{r}, b_1, \ldots, b_n) = 0$ that best approximates the behavior of (5.40) at t_0, and we have

$$F(\mathbf{r}(t_0), b_1, \ldots, b_n) = 0. \tag{5.42}$$

But as a practical matter it is not convenient to use such a limit passage to obtain the needed curve. It would be better to have conditions in which only those characteristics of the curves at t_0 are involved. Since we suppose the above limit passage is well-defined we can take the ordered set of points t_k:

$$t_0 < t_1 < \cdots < t_{n-1}.$$

Let $a_1, \ldots, a_n$ be a solution to (5.41) in this case, and consider the function

$$f(t) = F(\mathbf{r}(t), a_1, \ldots, a_n)$$

of the single variable t. This function vanishes at $t = t_k$ for $k = 0, \ldots, n-1$. By Rolle's theorem there exists $t'_k \in [t_{k-1}, t_k]$, $k = 1, \ldots, n-1$, such that $f'(t'_k) = 0$. During the limit passage all the t'_k also tend to t_0 so we have

$$\left. \frac{d}{dt} F(\mathbf{r}(t), b_1, \ldots, b_n) \right|_{t=t_0} = 0.$$

Similarly, the function

$$f'(t) = F_t(\mathbf{r}(t), a_1, \ldots, a_n)$$

takes $n-1$ zeroes at the points $t'_1 < \cdots < t'_{n-1}$. So by Rolle's theorem there is a point t''_k on each segment $[t'_{k-1}, t'_k]$ such that $f''(t''_k) = 0$. After the main limit passage we will have from this

$$\frac{d^2}{dt^2}F(\mathbf{r}(t), b_1, \ldots, b_n)\bigg|_{t=t_0} = 0.$$

Repeating this procedure for each derivative up to order $n-1$, we obtain the conditions

$$\frac{d^k}{dt^k}F(\mathbf{r}(t), b_1, \ldots, b_n)\bigg|_{t=t_0} = 0, \qquad k \le n-1.$$

These, taken together with (5.42), constitute n conditions that should be fulfilled by the needed curve from the family Φ_n. We define the *order of contact* between this curve and (5.40) as the number $n-1$.

Contact of a curve with a circle; evolutes

Let us return to our previous problem and apply these conditions to approximate a curve

$$\mathbf{r} = \mathbf{i}_1 x(t) + \mathbf{i}_2 y(t)$$

at a point $t = t_0$ by a circle of the family (5.39). Our three free parameters x_0, y_0, R should satisfy the system

$$\begin{aligned} (x(t_0) - x_0)^2 + (y(t_0) - y_0)^2 &= R^2, \\ 2(x(t_0) - x_0)x'(t_0) + 2(y(t_0) - y_0)y'(t_0) &= 0, \\ 2x'(t_0)^2 + 2(x(t_0) - x_0)x''(t_0) + 2y'(t_0)^2 + 2(y(t_0) - y_0)y''(t_0) &= 0. \end{aligned}$$

In particular, solution of these yields

$$R^2 = \frac{\{[x'(t_0)]^2 + [y'(t_0)]^2\}^3}{[x'(t_0)y''(t_0) - x''(t_0)y'(t_0)]^2}.$$

We thus see that the curvature $1/R$ of the circle coincides with the curvature k_1 of the curve. As the formulas for x_0 and y_0 are cumbersome, it is worthwhile to recast the problem in vector notation.

Let $\mathbf{r}_0$ locate the center of the contact circle

$$(\mathbf{r} - \mathbf{r}_0)^2 = R^2.$$

Let the curve be given in natural parameterization as $\mathbf{r} = \mathbf{r}(s)$. To solve this problem of second-order contact we define $F(s) = (\mathbf{r}(s) - \mathbf{r}_0)^2 - R^2$

and have

$$F(s_0) = 0, \qquad F'(s_0) = 0, \qquad F''(s_0) = 0,$$

or

$$\begin{aligned}(\mathbf{r}(s_0) - \mathbf{r}_0)^2 - R^2 &= 0,\\ 2(\mathbf{r}(s_0) - \mathbf{r}_0) \cdot \boldsymbol{\tau}(s_0) &= 0,\\ 2 + 2(\mathbf{r}(s_0) - \mathbf{r}_0) \cdot \boldsymbol{\nu}(s_0)k_1 &= 0.\end{aligned}$$

By the second equation the vector $\mathbf{r}(s_0) - \mathbf{r}_0$ is orthogonal to the tangent $\boldsymbol{\tau}(s_0)$, hence is directed along the normal $\boldsymbol{\nu}(s_0)$ and we have

$$(\mathbf{r}(s_0) - \mathbf{r}_0) \cdot \boldsymbol{\nu}(s_0) = -|\mathbf{r}(s_0) - \mathbf{r}_0| = -R.$$

This and the third equation yield $1 - k_1 R = 0$, hence $R = 1/k_1$ as stated above.

The locus of centers of all contact circles for a given curve is called the *evolute* of the curve. The equation of the evolute is

$$\boldsymbol{\rho}(t) = \mathbf{r}(t) + R\boldsymbol{\nu}(t), \qquad R = 1/k_1$$

or in parametric Cartesian form ($\boldsymbol{\rho} = (\xi, \eta)$)

$$\xi = x - y'\frac{x'^2 + y'^2}{x'y'' - x''y'}, \qquad \eta = y + x'\frac{x'^2 + y'^2}{x'y'' - x''y'}.$$

Contact of nth order between a curve and a surface

Quite similarly we can solve a problem of nth-order contact between a space curve $\mathbf{r} = \mathbf{r}(t)$ at $t = t_0$ and a surface from an $n+1$ parameter family of surfaces given implicitly by

$$F(\mathbf{r}, a_1, \ldots, a_{n+1}) = 0.$$

Everything from the previous pages should be repeated word for word. First we choose a surface from the family in such a way that it coincides with the curve in $n+1$ points close to t_0. This gives $n+1$ equations:

$$\begin{aligned}F(\mathbf{r}(t_0), a_1, \ldots, a_{n+1}) &= 0,\\ F(\mathbf{r}(t_1), a_1, \ldots, a_{n+1}) &= 0,\\ &\vdots\\ F(\mathbf{r}(t_{n-1}), a_1, \ldots, a_{n+1}) &= 0.\end{aligned}$$

Our previous reasoning carries through and we obtain

$$\left.\frac{d^k}{dt^k}F(\mathbf{r}(t), b_1, b_2, \ldots, b_{n+1})\right|_{t=t_0} = 0, \qquad k = 0, 1, \ldots, n,$$

as the equations that should hold at the point of nth-order contact.

We have introduced the osculating plane to a curve at a point A as the plane through A that contains $\boldsymbol{\tau}$ and $\boldsymbol{\nu}$. Note that it could also be defined as a surface of second-order contact. The result is the same.

The reader should consider how to apply these considerations to the problem of nth-order contact between a given surface and a surface from a many-parameter family of surfaces.

Exercise 5.36 Treat the problem of third-order contact between a space curve and a sphere. Show that denoting $R = 1/k_1$ and ρ the radius of the sphere we get (in natural parameterization)

$$\rho^2 = R^2 + R'^2/k_2^2.$$

Also show that the center of the contact sphere lies on the straight line through the center of principal curvature that is parallel to the binormal at the point of the curve.

5.8 Osculating Paraboloid

When considering the structure of a surface at a point, it is often helpful to approximate the surface using another surface whose behavior is more easily visualized. A spherical surface would be insufficient for this purpose because it has the same normal curvature in all directions. We can, however, use the osculating paraboloid introduced in (5.30). Let us reconsider this paraboloid from the point of view of local approximation.

Let O be a fixed point of a surface, and assume the surface is sufficiently smooth at O. At O we determine the osculating plane and introduce a Cartesian frame $(\mathbf{i}_1, \mathbf{i}_2, \mathbf{n})$, where $(\mathbf{i}_1, \mathbf{i}_2)$ is a Cartesian frame with origin O on the osculating plane and $\mathbf{n}$ is normal to both the surface and the osculating plane. For a smooth surface Cartesian coordinates (x, y) can play the role of surface coordinates since they uniquely define any point of the surface at O. The surface in the vicinity of O can be described by the equation $z = z(x, y)$. We suppose that $z(x, y)$ is twice continuously differentiable near O. In the neighborhood of $(0, 0)$ we can use the Taylor expansion of $z(x, y)$, which is

$$\begin{aligned} z(x,y) &= z(0,0) + z_x(0,0)x + z_y(0,0)y + \\ &+ \frac{1}{2}\left(z_{xx}x^2 + 2z_{xy}xy + z_{yy}y^2\right) + o(x^2+y^2) \end{aligned}$$

where the indices x, y indicate that we take partial derivatives with respect to x, y, respectively (and in this section, evaluated at the point $(0,0)$). Since the coordinates (x, y) are in the osculating plane, the partial derivatives

$$z_x(0,0) = 0, \qquad z_y(0,0) = 0.$$

By choice of the origin, $z(0,0) = 0$. Thus the Taylor expansion is

$$z(x,y) = \frac{1}{2}\left(z_{xx}x^2 + 2z_{xy}xy + z_{yy}y^2\right) + o(x^2+y^2). \tag{5.43}$$

Let us consider the paraboloid

$$z(x,y) = \frac{1}{2}\left(z_{xx}x^2 + 2z_{xy}xy + z_{yy}y^2\right). \tag{5.44}$$

The difference between the surfaces described by equations (5.43) and (5.44) is small (as indicated by the o term). This allows us to show that it is an osculating paraboloid that approximates the behavior of the surface at point O. A change of coordinate frames can show that the paraboloid (5.44) coincides with (5.30). Its coefficients are L, M, N at the point O.

Let us note that the second fundamental form of the initial surface at a point coincides with that of the osculating paraboloid, and the situation is the same with the normal curvatures. We denote

$$r = z_{xx}(0,0), \qquad s = z_{xy}(0,0), \qquad t = z_{yy}(0,0),$$

so that the equation of the osculating paraboloid is

$$z = \frac{1}{2}\left(rx^2 + 2sxy + ty^2\right).$$

Let us draw the projection onto the xy-coordinate plane of the cross-sections of the osculating paraboloid by the two planes $z = \pm h$, $h > 0$. This is the curve defined by

$$\frac{1}{2}\left|rx^2 + 2sxy + ty^2\right| = h. \tag{5.45}$$

The curve is an ellipse when the paraboloid is elliptical ($rt - s^2 > 0$) or a hyperbola when it is hyperbolic ($rt - s^2 < 0$), or a family of straight lines when it is parabolic ($rt - s^2 = 0$). It can be shown that the radius of normal curvature of the surface in the direction $x : y$ is proportional to the squared distance from the origin to the point of the curve (5.45) taken in the same direction. This curve is called the *Dupin indicatrix.*

The Dupin indicatrix is a curve of second order on the plane, and thus it has some special directions (axes) and special values characterizing the curve. The special directions are known as *principal directions.* In the next section we consider this from another vantage point.

5.9 The Principal Curvatures of a Surface

Let us discuss in more detail the properties of a surface connected with its second fundamental form. We now denote the coordinates without indices, using $u = u^1, v = u^2$. We also use subscripts u, v to indicate partial differentiation:

$$\mathbf{r}_1 = \mathbf{r}_u = \frac{\partial \mathbf{r}}{\partial u}, \qquad \mathbf{r}_2 = \mathbf{r}_v = \frac{\partial \mathbf{r}}{\partial v}, \qquad \mathbf{n}_u = \frac{\partial \mathbf{n}}{\partial u}, \qquad \mathbf{n}_v = \frac{\partial \mathbf{n}}{\partial v}.$$

Hence the second fundamental form is

$$L\, du^2 + 2M\, du\, dv + N\, dv^2 = -d\mathbf{r} \cdot d\mathbf{n}$$

where

$$L = \mathbf{r}_{uu} \cdot \mathbf{n}, \qquad M = \mathbf{r}_{uv} \cdot \mathbf{n}, \qquad N = \mathbf{r}_{vv} \cdot \mathbf{n}.$$

Consider the first differential of $\mathbf{n}$ at a point P:

$$d\mathbf{n} = \mathbf{n}_u\, du + \mathbf{n}_v\, dv.$$

Since $\mathbf{n}$ is a unit vector its differential $d\mathbf{n}$ is orthogonal to $\mathbf{n}$ and thus lies in the osculating plane to the surface at P. We know that the differential $d\mathbf{r} = \mathbf{r}_u\, du + \mathbf{r}_v\, dv$ also lies in the plane osculating at P. Let us consider the relation between the differentials $d\mathbf{r}$ and $d\mathbf{n}$ with respect to the variables du, dv now considered as independent variables. It is clearly a linear correspondence $d\mathbf{r} \mapsto d\mathbf{n}$. Thus it defines a tensor $\mathbf{A}$ in two-dimensional space such that

$$d\mathbf{n} = \mathbf{A} \cdot d\mathbf{r}.$$

This tensor is completely defined by its values $\mathbf{n}_u = \mathbf{A} \cdot \mathbf{r}_u$ and $\mathbf{n}_v = \mathbf{A} \cdot \mathbf{r}_v$.

Lemma 5.1 *The tensor $\mathbf{A}$ is symmetric.*

Proof. It is enough to establish the equality

$$\mathbf{x}_1 \cdot (\mathbf{A} \cdot \mathbf{x}_2) = \mathbf{x}_2 \cdot (\mathbf{A} \cdot \mathbf{x}_1)$$

for a pair of linearly independent vectors $(\mathbf{x}_1, \mathbf{x}_2)$. To show symmetry of $\mathbf{A}$ consider

$$\mathbf{r}_u \cdot (\mathbf{A} \cdot \mathbf{r}_v) = \mathbf{r}_u \cdot \mathbf{n}_v.$$

Similarly

$$\mathbf{r}_v \cdot (\mathbf{A} \cdot \mathbf{r}_u) = \mathbf{r}_v \cdot \mathbf{n}_u.$$

The symmetry of $\mathbf{A}$ follows from the identity

$$\mathbf{r}_u \cdot \mathbf{n}_v = \mathbf{r}_v \cdot \mathbf{n}_u;$$

this is derived by differentiating the identity $\mathbf{n} \cdot \mathbf{r}_u = 0$ with respect to v, then differentiating $\mathbf{n} \cdot \mathbf{r}_v = 0$ with respect to u and eliminating the term containing $\mathbf{r}_{uv}$. □

We know that a symmetric tensor has real eigenvalues; in this case there are no more than two eigenvalues λ_1 and λ_2 to which there correspond mutually orthogonal eigenvectors $\mathbf{x}_1, \mathbf{x}_2$, respectively (if $\lambda_1 \neq \lambda_2$).

Let us find the equation for the eigenvalues. A vector in the osculating plane can be represented as $x\mathbf{r}_u + y\mathbf{r}_v$ since $(\mathbf{r}_u, \mathbf{r}_v)$ is a basis in it. By definition of $\mathbf{A}$ we have

$$\mathbf{A} \cdot (x\mathbf{r}_u + y\mathbf{r}_v) = x\mathbf{n}_u + y\mathbf{n}_v.$$

On the other hand $x\mathbf{r}_u + y\mathbf{r}_v \neq 0$ is an eigenvector if there exists λ such that

$$\mathbf{A} \cdot (x\mathbf{r}_u + y\mathbf{r}_v) = \lambda\, (x\mathbf{r}_u + y\mathbf{r}_v)\,.$$

Thus, for the same vector we get

$$x\mathbf{n}_u + y\mathbf{n}_v = \lambda\, (x\mathbf{r}_u + y\mathbf{r}_v)\,. \tag{5.46}$$

From this we see that an eigenvector is defined by the same condition as the principal directions of the previous section, since this equation means that we find the direction in which $d\mathbf{r}$ and $d\mathbf{n}$ are parallel.

From the last vector equation, let us derive scalar equations. For this, dot multiply (5.46) first by $\mathbf{r}_u$, then by $\mathbf{r}_v$. By (5.31) we have

$$L = -\mathbf{n}_u \cdot \mathbf{r}_u, \qquad M = -\mathbf{n}_u \cdot \mathbf{r}_v, \qquad N = -\mathbf{n}_v \cdot \mathbf{r}_v,$$

hence

$$-Lx - My = \lambda(Ex + Fy),$$
$$-Mx - Ny = \lambda(Fx + Gy).$$

Rewriting this as

$$(L + \lambda E)x + (M + \lambda F)y = 0,$$
$$(M + \lambda F)x + (N + \lambda G)y = 0, \tag{5.47}$$

we have a homogeneous linear system of algebraic equations. This system has nontrivial solutions when its determinant vanishes:

$$(EG - F^2)\lambda^2 - (2MF - EN - LG)\lambda + (LN - M^2) = 0. \tag{5.48}$$

In the general case, there are two roots λ_1 and λ_2, to which there correspond two directions previously called the principal directions. They can be found by elimination of λ from (5.47):

$$\begin{vmatrix} -x^2 & xy & -y^2 \\ E & F & G \\ L & M & N \end{vmatrix} = 0.$$

By this we define two mutually orthogonal directions $x : y$. If $\lambda_1 = \lambda_2$, then all the directions are principal so we can choose any two mutually orthogonal directions and regard them as principal.

Now we would like to consider another question that will bring us to the same equations. It is the question of finding the extreme normal curvatures at a point of a surface. We have established that in the direction $x : y$ the normal curvature of a surface is given by the formula

$$k = \frac{Lx^2 + 2Mxy + Ny^2}{Ex^2 + 2Fxy + Gy^2}.$$

Because of the second-order homogeneity in x, y of the numerator and denominator we can reformulate the problem of finding extreme curvatures as the problem of finding extremal points of the function

$$Lx^2 + 2Mxy + Ny^2$$

under the restriction

$$Ex^2 + 2Fxy + Gy^2 = 1.$$

Applying the theory of Lagrange multipliers to this we should find the extreme points of the function

$$Lx^2 + 2Mxy + Ny^2 - k(Ex^2 + 2Fxy + Gy^2)$$

which leads to the equations

$$\begin{aligned}(L - kE)x + (M - kF)y &= 0\\ (M - kF)x + (N - kG)y &= 0.\end{aligned} \tag{5.49}$$

We see that the systems (5.47) and (5.49) coincide if we put $k = -\lambda$, which means that the principal directions and the directions found here are the same. Moreover, it is seen that λ_1 and λ_2 found as the eigenvalues of the tensor **A** give the extreme curvatures of the surface at a point that $k_1 = -\lambda_1$ and $k_2 = -\lambda_2$. Remembering that they are the roots of the polynomial (5.48) and using the Viète theorem we get

$$k_1 + k_2 = \frac{EN - 2FM + GL}{EG - F^2}, \qquad k_1 k_2 = \frac{LN - M^2}{EG - F^2}.$$

We have met these expressions that were called $K = k_1 k_2$, Gaussian curvature, and $H = (k_1 + k_2)/2$, the mean curvature.

We see that both are expressed through the coefficients of the first and second fundamental forms of the surface. There is a famous theorem due to Gauss that K can be expressed only in terms of the coefficients of the first fundamental form.

When the principal curvatures and their directions are known, the Euler formula gives the normal curvature at any direction of the curve which composes the angle ϕ with the first principal direction at the same point, corresponding to k_1:

$$k_\phi = k_1 \cos\phi + k_2 \sin\phi.$$

The principal directions of a surface define the *lines of curvature* of the surface. A line in the surface is called a line of curvature if at each point its tangent is directed along one of the principal directions at this point of the surface.

At each point there are two lines of curvature through the point. Denoting their directions by $du : dv$ and $\delta u : \delta v$ we have two equations, the first of which means orthogonality of the directions and the second their conjugation:

$$E\,du\,\delta u + F(du\,\delta v + dv\,\delta u) + G\,dv\,\delta v = 0,$$
$$L\,du\,\delta u + M(du\,\delta v + dv\,\delta u) + N\,dv\,\delta v = 0.$$

It is convenient to take the family of the lines of curvature of the surface as the coordinate lines of the surface. For these curvilinear coordinates $F = M = 0$.

Two surfaces are called *isometric* if there is a one-to-one correspondence between the points of the surfaces such that corresponding curves in the surfaces have the same length.

A planar surface and a cylindrical surface provide an example of isometric surfaces since we can develop the cylindrical surface over the plane. The correspondence between points is defined by coincidence of the points in this developing.

We can refer to local isometry of a surface at point A_1 to another surface at point B_2 if there are neighborhoods of points on the surfaces that are isometric.

Two smooth surfaces are locally isometric if and only if there are parameterizations of the surfaces at corresponding points such that the coefficients of the first principal forms of the surfaces coincide:

$$E_1 = E_2, \qquad F_1 = F_2, \qquad G_1 = G_2.$$

An interesting class of surfaces is the class of surfaces that is locally isometric to the plane at each point. Such a surface is called *developable.*

It turns out that a surface is developable if and only if its Gaussian curvature is zero at each point.

Surfaces with zero Gaussian curvature appear once more in the problem of finding spatial surfaces of minimal area that have given boundaries. This is a popular physics problem since it describes the shape assumed by a soap film on a wire frame. The energy of such a film is proportional to its area, and thus the form actually taken by such a film is the surface of minimum area.

5.10 Surfaces of Revolution

Surfaces of revolution are quite frequent in practice. Suppose that the surface is formed by rotation of the profile curve

$$x = \phi(u), \qquad z = \psi(u), \tag{5.50}$$

in the xz-plane about z-axis (Fig. 5.3). When we fix u we get a circle having center on the z axis; it is called a *parallel*. To define a point on the parallel we introduce the angle of rotation v from the xz-plane. When we fix v we get a *meridian*, a curve congruent to the initial curve (5.50).

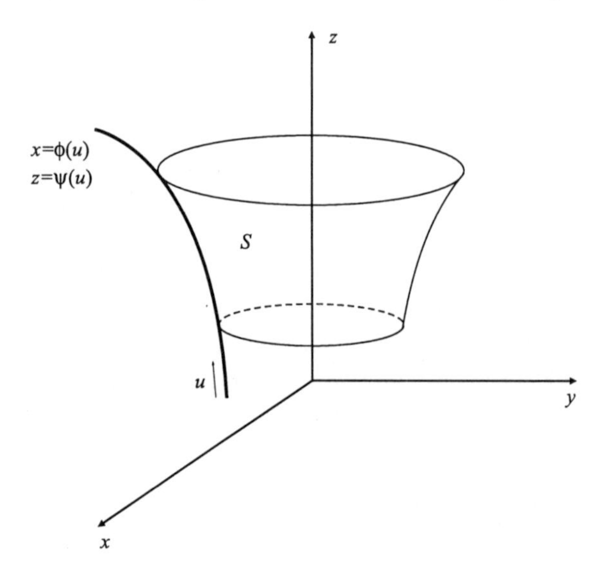

Fig. 5.3 Surface of revolution about the z-axis.

It is easy to see that the equations of the surface of revolution corresponding (5.50) are

$$x = \phi(u)\cos v, \qquad y = \phi(u)\sin v, \qquad z = \psi(u).$$

Let us find the coefficients of the first fundamental form of the surface:

$$\begin{aligned}
E &= x_u^2 + y_u^2 + z_u^2 = (\phi' \cos v)^2 + (\phi' \sin v)^2 + \psi'^2 = \phi'^2 + \psi'^2, \\
F &= x_u x_v + y_u y_v + z_u z_v = (\phi' \cos v)(-\phi \sin v) + (\phi' \sin v)(\phi \cos v) = 0, \\
G &= x_v^2 + y_v^2 + z_v^2 = (-\phi \sin v)^2 + (\phi \cos v)^2 = \phi^2.
\end{aligned}$$

Note that $F = 0$ means the orthogonality of the parameterization net. Thus the first fundamental form is

$$(ds)^2 = \left(\phi'^2 + \psi'^2\right) du^2 + \phi^2 \, dv^2.$$

For the components of the second fundamental form we have

$$L = \frac{\mathbf{r}_{uu} \cdot (\mathbf{r}_u \times \mathbf{r}_v)}{|\mathbf{r}_u \times \mathbf{r}_v|} = \frac{\begin{vmatrix} x_{uu} & y_{uu} & z_{uu} \\ x_u & y_u & z_u \\ x_v & y_v & z_v \end{vmatrix}}{\sqrt{EG - F^2}} = \frac{\psi''\phi' - \phi''\psi'}{\sqrt{\phi'^2 + \psi'^2}},$$

$$M = \frac{\mathbf{r}_{uv} \cdot (\mathbf{r}_u \times \mathbf{r}_v)}{|\mathbf{r}_u \times \mathbf{r}_v|} = \frac{\begin{vmatrix} x_{uv} & y_{uv} & z_{uv} \\ x_u & y_u & z_u \\ x_v & y_v & z_v \end{vmatrix}}{\sqrt{EG - F^2}} = 0,$$

and

$$N = \frac{\mathbf{r}_{vv} \cdot (\mathbf{r}_u \times \mathbf{r}_v)}{|\mathbf{r}_u \times \mathbf{r}_v|} = \frac{\begin{vmatrix} x_{vv} & y_{vv} & z_{vv} \\ x_u & y_u & z_u \\ x_v & y_v & z_v \end{vmatrix}}{\sqrt{EG - F^2}} = \frac{\psi'\phi}{\sqrt{\phi'^2 + \psi'^2}}.$$

Thus the second fundamental form is

$$-d\mathbf{n} \cdot d\mathbf{r} = \frac{\psi''\phi' - \phi''\psi'}{\sqrt{\phi'^2 + \psi'^2}} \, du^2 + \frac{\psi'\phi}{\sqrt{\phi'^2 + \psi'^2}} \, dv^2.$$

We see that $M = 0$, which means that the coordinate lines are conjugate.[6] Thus the coordinate lines of a surface of revolution are the lines of curvature.

Exercise 5.37 Find the principal curvatures of the surface of revolution.

Exercise 5.38 Find the first and second fundamental forms for (a) the plane, and (b) the sphere.

Exercise 5.39 Find the first and second fundamental forms for each of the following paraboloids: (a) $z = x^2 + y^2$, (b) $z = x^2 - y^2$, (c) $z = x^2$. Find H and K for each of these.

[6]To any symmetric quadratic form there corresponds something like an inner product of vectors, and hence something akin to orthogonality. Conjugate directions at a point are defined as the directions on the surface $du : dv$ and $\delta u : \delta v$ for which $L \, du\delta u + M(du\delta v + dv\delta u) + N \, dv\delta v = 0$. Conjugate coordinate lines are those which are conjugate at each point.

5.11 Natural Equations of a Curve

Suppose we are given functions $k_1(s)$ and $k_2(s)$ of a natural parameter s, continuous on a segment $[s_0, s_1]$, and that $k_1(s)$ is a positive function. We are interested in whether there exists a space curve that has principal curvature $k_1(s)$ and torsion $k_2(s)$. Geometrical considerations show that if such a curve exists then it is uniquely defined up to rigid motions. This means that if we take two such curves, place them in space, and then shift and rotate one of them so that their initial points and moving trihedra at these points coincide, then all points of the curves coincide.

Now we will show that such a curve exists. Thus the form of the curve is defined by $k_1(s)$ and $k_2(s)$ uniquely; this prompts us to call the two functions $k_1 = k_1(s)$ and $k_2 = k_2(s)$ the *natural equations* of the curve.

Let us demonstrate that from $k_1(s)$ and $k_2(s)$ we can find the needed curve. For this, we consider a vector system of differential equations that mimics the formula for a tangent vector and the Serret–Frenet equations:

$$\begin{aligned}
\frac{d\mathbf{r}}{ds} &= \mathbf{x}, \\
\frac{d\mathbf{x}}{ds} &= k_1(s)\mathbf{y}, \\
\frac{d\mathbf{y}}{ds} &= -k_1(s)\mathbf{x} - k_2(s)\mathbf{z}, \\
\frac{d\mathbf{z}}{ds} &= k_2(s)\mathbf{y}.
\end{aligned} \tag{5.51}$$

When written in Cartesian component form this is a system of 12 linear ordinary differential equations. ODE theory states that if we define the initial values for all the unknowns (the Cauchy problem), then we will have a unique solution to this system.

We can choose arbitrary initial conditions for $\mathbf{r}(s_0)$. The initial values $\mathbf{x}(s_0)$, $\mathbf{y}(s_0)$, $\mathbf{z}(s_0)$ must constitute an arbitrary right-handed orthonormal trihedron $(\mathbf{x}_0, \mathbf{y}_0, \mathbf{z}_0)$.

Thus on $[s_0, s_1]$ there exists a unique solution $(\mathbf{r}(s), \mathbf{x}(s), \mathbf{y}(s), \mathbf{z}(s))$ of the equations (5.51), satisfying some fixed initial conditions. It can be shown that because of skew symmetry of the matrix of the three last equations of (5.51),

$$\begin{vmatrix} 0 & k_1(s) & 0 \\ -k_1(s) & 0 & -k_2(s) \\ 0 & k_2(s) & 0 \end{vmatrix} = 0,$$

the vectors $(\mathbf{x}(s), \mathbf{y}(s), \mathbf{z}(s))$ constitute an orthonormal frame of the same orientation as the initial one on the whole segment $[s_0, s_1]$.

We can treat $\mathbf{r} = \mathbf{r}(s)$ as the equation of the needed curve in natural parameterization. Then the first of the equations (5.51) states that $\mathbf{x}(s)$ is its unit tangent. Comparing the equation $d\mathbf{x}/ds = k_1(s)\mathbf{y}$ with the first of the Serret–Frenet equations for a curve $\mathbf{r} = \mathbf{r}(s)$, we see that $\mathbf{y}(s)$ is its principal normal and $k_1(s)$ is its principal curvature. Similarly we find that $\mathbf{z}(s)$ is the binormal of the curve and $k_2(s)$ is its torsion. This completes the necessary reasoning.

For a plane curve the natural equations reduce to a single equation for the curvature. In the plane we associate the curvature with an algebraic sign. The condition of positivity of $k(s)$ is not necessary.

Natural equation of a curve in the plane

A plane curve has zero torsion. Let us consider how to reconstruct a curve if its curvature $k(s)$ is a given function of the length parameter s. We fix the initial point of the curve, corresponding to $s = s_0$, by the equation

$$\mathbf{r}(s_0) = \mathbf{r}_0. \tag{5.52}$$

We must also fix the direction of the curve at this point. Here it is useful to introduce the angle $\phi = \phi(s)$, measured between the unit tangent to the curve and the vector $\mathbf{i}$ where $\{\mathbf{i}, \mathbf{j}\}$ is an orthonormal basis in the plane. In this way the unit tangent is given by

$$\boldsymbol{\tau} = \mathbf{i}\cos\phi + \mathbf{j}\sin\phi.$$

So to define the initial direction of the curve we introduce

$$\phi(s_0) = \phi_0. \tag{5.53}$$

The equation relating $\boldsymbol{\tau}$ and $\boldsymbol{\nu}$, which is $d\boldsymbol{\tau}/ds = k(s)\boldsymbol{\nu}$, becomes

$$(-\mathbf{i}\sin\phi + \mathbf{j}\cos\phi)\frac{d\phi}{ds} = k(s)\boldsymbol{\nu}.$$

The vector $-\mathbf{i}\sin\phi + \mathbf{j}\cos\phi$ has unit magnitude and is orthogonal to $\boldsymbol{\tau}$. Hence it is parallel to $\boldsymbol{\nu}$, and we conclude that

$$|k(s)| = \left|\frac{d\phi}{ds}\right|.$$

Taking

$$k(s) = \frac{d\phi}{ds},$$

we define a sign of $k(s)$ in the standard manner where $k(s)$ is positive for points at which the curve is concave upwards.

We can integrate the last equation and obtain

$$\phi(s) - \phi_0 = \int_{s_0}^{s} k(t)\, dt.$$

Knowing $\phi(s)$ we can integrate the equation for the unit tangent, which is $d\mathbf{r}/ds = \boldsymbol{\tau}$, rewritten as

$$\frac{d\mathbf{r}}{ds} = \mathbf{i} \cos \phi(s) + \mathbf{j} \sin \phi(s).$$

Now integration with respect to s gives us

$$\mathbf{r}(s) - \mathbf{r}_0 = \int_{s_0}^{s} [\mathbf{i} \cos \phi(t) + \mathbf{j} \sin \phi(t)]\, dt. \tag{5.54}$$

We see that by (5.54) the function $k(s)$ and the initial values (5.52) and (5.53) uniquely defined the needed curve $\mathbf{r} = \mathbf{r}(s)$. Changing the initial values of the curve we get a family of curves with the same $k(s)$ such that all curves have the same shape but different positions with respect to the coordinate axes.

Exercise 5.40 Given the curvature $k(s) = (as)^{-1}$ of a plane curve, find the curve.

5.12 A Word About Rigor

The main goal of this book has been the presentation of those tools and formulas of tensor analysis that are needed for applications. Our approach has been quite typical of engineering books; we seldom offered clear statements of the assumptions that guarantee validity of a formula, supposing instead that in applications all the functions, curves, surfaces, etc., should be sufficiently smooth for our purposes. This approach is probably best for any practitioner who must simply get his or her hands on a formula. However, in physics we see surfaces that do not necessarily bound real-world bodies; purely mathematical surfaces can occur, such as describe the energy of a two-parameter mechanical system. Such a surface can be quite complex, and a physicist may be largely interested in its singular points since these represent the states at which the system changes its behavior crucially. A reader interested in such applications would do well to study a more sophisticated treatment where each major result is stated as the theorem under the weakest possible hypotheses. Many such treatments employ much more advanced mathematical tools, including the branch of mathematics known as topology.

However, the reader should be aware that even on the elementary level of our treatment there are some questions that require additional explanation. How, for example, should we define the length of a curve or the area of a surface?

Early in our education we learn how to measure the length of a segment or of a more complex set on a straight line. We also learn how to define the circumference of a circle. For this we inscribe an equilateral triangle and calculate its perimeter. Doubling the number of sides of the inscribed triangle, we get an inscribed polygon whose perimeter approximates the circumference. Doubling the number of sides to infinity and calculating the limit of their perimeters, we define this limit as the needed length. Of course, a limit passage of this type is used when we derive the formula

$$\int_a^b |\mathbf{r}'(t)|\, dt \tag{5.55}$$

for the length of a curve. Here we divide the segment $[a, b]$ into small pieces by the points $t_0 = a, t_1, \ldots, t_n = b$, and draw the vector

$$\Delta\mathbf{r}(t_i + 1) = \mathbf{r}(t_{i+1}) - \mathbf{r}(t_i).$$

In this way we "inscribe" a polygon into the curve. For a continuously differentiable vector function $\mathbf{r}(t)$, the length of a side of a polygon can be found by the mean value theorem:

$$|\mathbf{r}(t_{i+1}) - \mathbf{r}(t_i)| = |\mathbf{r}'(\xi_{i+1})|(t_{i+1} - t_i),$$

where ξ_{i+1} is a point in $[t_{i+1} - t_i]$. Thus the perimeter of the inscribed polygon that approximates the length of the curve is

$$\sum_{i=1}^{n} |\mathbf{r}'(\xi_{i+1})|(t_{i+1} - t_i).$$

We obtain (5.55) from a limit passage in which the number of partition points tends to infinity while the maximum length of a partition segment tends to zero.

Although everything might seem fine with this, we should note that underlying the process is the notion of approximating a small piece of the curve by a many-sided polygon. Our "intuition" tells us that the smaller its sides, the closer is the polygon to the curve, and thus its perimeter should approximate the curve length more and more closely. But in elementary geometry one may "demonstrate" that the sum of the legs of any right triangle is equal to its hypotenuse. The construction is as follows. Suppose we are given a right triangle whose legs have lengths a and b. Let us form a curve that looks like the toothed edge of a saw, by placing along the hypotenuse of the given triangle a sequence of small triangles each similar to the given triangle (Fig. 5.4). It is clear that the sum of all the legs of the sawteeth does not depend on the number of teeth and equals $a + b$. But as the number of teeth tends to infinity the saw edge "coincides" with the hypotenuse of the original triangle. Hence the limiting length is equal to the length of the hypotenuse c, and we have $c = a + b$. So we come to understand that we cannot approximate a curve by a polygon in an absolutely arbitrary fashion. Similar remarks apply to the approximation of surfaces to calculate area.

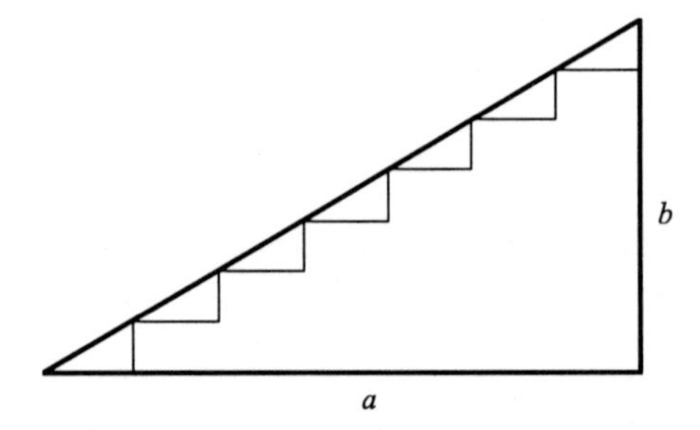

Fig. 5.4 Fallacious estimation of the length of a line.

These remarks should serve as a warning that to apply formulas correctly it is often necessary to understand the restrictions under which they were derived.

5.13 Conclusion

Differential geometry is a well developed branch of mathematics with many results and formulas. We have presented the main technical formulas that are used in applications. The interested reader could go on to study volumes devoted to the theory of curves, surfaces, manifolds, etc. But such an extensive undertaking falls outside the scope of this book.

Appendix A

Formulary

For convenience we list here the main tensor formulas obtained in each chapter. Note: the symbol $\forall$ denotes the universal quantifier (read as "for all" or "for every").

Chapter 2

Reciprocal (dual) basis

Kronecker delta symbol

$$\delta^i_j = \begin{cases} 1 & j = i \\ 0 & j \neq i \end{cases}$$

Definition of reciprocal basis

$$\mathbf{e}_j \cdot \mathbf{e}^i = \delta^i_j$$

Components of a vector

$$\mathbf{x} = x^i \mathbf{e}_i = x_i \mathbf{e}^i \qquad \begin{aligned} x^i &= \mathbf{x} \cdot \mathbf{e}^i \\ x_i &= \mathbf{x} \cdot \mathbf{e}_i \end{aligned}$$

Relations between dual bases

$$\mathbf{e}^i = \frac{1}{V}(\mathbf{e}_j \times \mathbf{e}_k) \qquad \mathbf{e}_i = \frac{1}{V'}(\mathbf{e}^j \times \mathbf{e}^k)$$

where

$$(i,j,k) = (1,2,3) \text{ or } (2,3,1) \text{ or } (3,1,2)$$

and

$$V = \mathbf{e}_1 \cdot (\mathbf{e}_2 \times \mathbf{e}_3) \qquad V' = 1/V$$
$$V' = \mathbf{e}^1 \cdot (\mathbf{e}^2 \times \mathbf{e}^3)$$

Metric coefficients

$$g^{jq} = \mathbf{e}^j \cdot \mathbf{e}^q \qquad g_{ij}g^{jk} = \delta_i^k$$
$$g_{ip} = \mathbf{e}_i \cdot \mathbf{e}_p$$

In Cartesian frames,

$$g_{ij} = \delta_i^j \qquad g^{ij} = \delta_j^i$$

Dot products in mixed and unmixed bases

$$\mathbf{a} \cdot \mathbf{b} = a^i b^j g_{ij} = a_i b_j g^{ij} = a^i b_i = a_i b^i$$

Raising and lowering of indices

$$x_j = x^i g_{ij} \qquad x^i = x_j g^{ij}$$

Frame transformation

Equations of transformation

$$\mathbf{e}_i = A_i^j \tilde{\mathbf{e}}_j \qquad A_i^j = \mathbf{e}_i \cdot \tilde{\mathbf{e}}^j$$
$$\tilde{\mathbf{e}}_i = \tilde{A}_i^j \mathbf{e}_j \qquad \tilde{A}_i^j = \tilde{\mathbf{e}}_i \cdot \mathbf{e}^j$$

where

$$\tilde{A}_i^j A_j^k = A_i^j \tilde{A}_j^k = \delta_i^k$$

Vector components and transformation laws

$$\mathbf{x} = x^i \mathbf{e}_i = x_i \mathbf{e}^i = \tilde{x}^i \tilde{\mathbf{e}}_i = \tilde{x}_i \tilde{\mathbf{e}}^i$$

and

$$\tilde{x}^i = A^i_j x^j \qquad x^i = \tilde{A}^i_j \tilde{x}^j$$
$$\tilde{x}_i = \tilde{A}^j_i x_j \qquad x_i = A^j_i \tilde{x}_j$$

Miscellaneous

Permutation (Levi–Civita) symbol

$$\epsilon_{ijk} = \mathbf{e}_i \cdot (\mathbf{e}_j \times \mathbf{e}_k) = \begin{cases} +V & (i,j,k) \text{ an even permutation of } (1,2,3) \\ -V & (i,j,k) \text{ an odd permutation of } (1,2,3) \\ 0 & \text{two or more indices equal} \end{cases}$$

$$\epsilon^{ijk} = \mathbf{e}^i \cdot (\mathbf{e}^j \times \mathbf{e}^k) = \begin{cases} +V' & (i,j,k) \text{ an even permutation of } (1,2,3) \\ -V' & (i,j,k) \text{ an odd permutation of } (1,2,3) \\ 0 & \text{two or more indices equal} \end{cases}$$

Useful identities

$$\epsilon_{ijk}\epsilon^{pqr} = \begin{vmatrix} \delta^p_i & \delta^q_i & \delta^r_i \\ \delta^p_j & \delta^q_j & \delta^r_j \\ \delta^p_k & \delta^q_k & \delta^r_k \end{vmatrix} \qquad \epsilon_{ijk}\epsilon^{pqk} = \delta^p_i\delta^q_j - \delta^q_i\delta^p_j$$

Determinant of Gram matrix

$$V^2 = \det[g_{ij}]$$

Cross product

$$\mathbf{a} \times \mathbf{b} = \mathbf{e}^i \epsilon_{ijk} a^j b^k = \mathbf{e}_i \epsilon^{ijk} a_j b_k$$

Chapter 3

Dyad product

Properties

$$
\begin{aligned}
(\lambda\mathbf{a}) \otimes \mathbf{b} &= \mathbf{a} \otimes (\lambda\mathbf{b}) = \lambda(\mathbf{a} \otimes \mathbf{b}) \\
(\mathbf{a}+\mathbf{b}) \otimes \mathbf{c} &= \mathbf{a} \otimes \mathbf{c} + \mathbf{b} \otimes \mathbf{c} \\
\mathbf{a} \otimes (\mathbf{b}+\mathbf{c}) &= \mathbf{a} \otimes \mathbf{b} + \mathbf{a} \otimes \mathbf{c}
\end{aligned}
$$

Dot products of dyad with vector

$$
\begin{aligned}
\mathbf{ab} \cdot \mathbf{c} &= (\mathbf{b} \cdot \mathbf{c})\mathbf{a} \\
\mathbf{c} \cdot (\mathbf{ab}) &= (\mathbf{c} \cdot \mathbf{a})\mathbf{b}
\end{aligned}
$$

Tensors from operator viewpoint

Equality of tensors

$$
\mathbf{A} = \mathbf{B} \iff \forall \mathbf{x},\ \mathbf{A} \cdot \mathbf{x} = \mathbf{B} \cdot \mathbf{x}
$$

Components

$$
\begin{aligned}
a^{ij} &= \mathbf{e}^i \cdot \mathbf{A} \cdot \mathbf{e}^j \\
a_{ij} &= \mathbf{e}_i \cdot \mathbf{A} \cdot \mathbf{e}_j \\
a^i_{\cdot j} &= \mathbf{e}^i \cdot \mathbf{A} \cdot \mathbf{e}_j \\
a_i^{\cdot j} &= \mathbf{e}_i \cdot \mathbf{A} \cdot \mathbf{e}^j
\end{aligned}
$$

Definition of sum $\mathbf{A} + \mathbf{B}$

$$
\forall \mathbf{x},\ (\mathbf{A} + \mathbf{B}) \cdot \mathbf{x} = \mathbf{A} \cdot \mathbf{x} + \mathbf{B} \cdot \mathbf{x}
$$

Definition of scalar multiple $c\mathbf{A}$

$$
\forall \mathbf{x},\ (c\mathbf{A}) \cdot \mathbf{x} = c(\mathbf{A} \cdot \mathbf{x})
$$

Definition of dot product $\mathbf{A} \cdot \mathbf{B}$

$$(\mathbf{A} \cdot \mathbf{B}) \cdot \mathbf{x} \equiv \mathbf{A} \cdot (\mathbf{B} \cdot \mathbf{x})$$

Definition of pre-multiplication $\mathbf{y} \cdot \mathbf{A}$

$$\forall \mathbf{x},\ (\mathbf{y} \cdot \mathbf{A}) \cdot \mathbf{x} = \mathbf{y} \cdot (\mathbf{A} \cdot \mathbf{x})$$

Definition of unit tensor $\mathbf{E}$

$$\forall \mathbf{x},\ \mathbf{E} \cdot \mathbf{x} = \mathbf{x} \cdot \mathbf{E} = \mathbf{x}$$

Unit tensor components

$$\mathbf{E} = \mathbf{e}^i \mathbf{e}_i = \mathbf{e}_j \mathbf{e}^j = g_{ij} \mathbf{e}^i \mathbf{e}^j = g^{ij} \mathbf{e}_i \mathbf{e}_j$$

Inverse tensor

$$\mathbf{A} \cdot \mathbf{A}^{-1} = \mathbf{E}$$

$$(\mathbf{A} \cdot \mathbf{B})^{-1} = \mathbf{B}^{-1} \cdot \mathbf{A}^{-1}$$

Nonsingular tensor $\mathbf{A}$

$$\mathbf{A} \cdot \mathbf{x} = 0 \implies \mathbf{x} = 0$$

Determinant of a tensor

$$\begin{aligned} \det \mathbf{A} &= \left| a_i^{\cdot j} \right| = \left| a^k_{\cdot m} \right| = \frac{1}{g} \left| a_{st} \right| = g \left| a^{pq} \right| \\ &= \frac{1}{6} \epsilon_{ijk} \epsilon^{mnp} a_m^{\cdot i} a_n^{\cdot j} a_p^{\cdot k} \end{aligned}$$

Dyadic components under transformation

Transformation to reciprocal basis

$$a_{km} = a^{ij} g_{ki} g_{jm}$$

More general transformation

$$\mathbf{e}_i = A_i^j \tilde{\mathbf{e}}_j \implies \tilde{a}^{ij} = a^{km} A_k^i A_m^j$$

$$\tilde{\mathbf{e}}_i = \tilde{A}_i^j \mathbf{e}_j \implies a^{ij} = \tilde{a}^{km} \tilde{A}_k^i \tilde{A}_m^j$$

$$A_j^k \tilde{A}_k^i = \delta_j^i$$

$$\begin{aligned} \mathbf{A} &= \tilde{a}^{ij} \tilde{\mathbf{e}}_i \tilde{\mathbf{e}}_j = \tilde{a}_{kl} \tilde{\mathbf{e}}^k \tilde{\mathbf{e}}^l = \tilde{a}_i^{\cdot j} \tilde{\mathbf{e}}^i \tilde{\mathbf{e}}_j = \tilde{a}_{\cdot l}^k \tilde{\mathbf{e}}_k \tilde{\mathbf{e}}^l \\ &= a^{ij} \mathbf{e}_i \mathbf{e}_j = a_{kl} \mathbf{e}^k \mathbf{e}^l = a_i^{\cdot j} \mathbf{e}^i \mathbf{e}_j = a_{\cdot l}^k \mathbf{e}_k \mathbf{e}^l \end{aligned}$$

where

$$\begin{aligned} \tilde{a}^{ij} &= A_k^i A_l^j a^{kl} & a^{ij} &= \tilde{A}_k^i \tilde{A}_l^j \tilde{a}^{kl} \\ \tilde{a}_{ij} &= \tilde{A}_i^k \tilde{A}_j^l a_{kl} & a_{ij} &= A_i^k A_j^l \tilde{a}_{kl} \\ \tilde{a}^i_{\cdot j} &= A_k^i \tilde{A}_j^l a^k_{\cdot l} & a^i_{\cdot j} &= \tilde{A}_k^i A_j^l \tilde{a}^k_{\cdot l} \\ \tilde{a}_i^{\cdot j} &= \tilde{A}_i^k A_l^j a_k^{\cdot l} & a_i^{\cdot j} &= A_i^k \tilde{A}_l^j \tilde{a}_k^{\cdot l} \end{aligned}$$

More dyadic operations

Dot product

$$\mathbf{ab} \cdot \mathbf{cd} = (\mathbf{b} \cdot \mathbf{c})\mathbf{ad}$$

$$\begin{aligned} \mathbf{A} \cdot (\lambda \mathbf{a} + \mu \mathbf{b}) &= \lambda \mathbf{A} \cdot \mathbf{a} + \mu \mathbf{A} \cdot \mathbf{b} \\ (\lambda \mathbf{A} + \mu \mathbf{B}) \cdot \mathbf{a} &= \lambda \mathbf{A} \cdot \mathbf{a} + \mu \mathbf{B} \cdot \mathbf{a} \end{aligned}$$

Double dot product

$$\mathbf{ab} \cdot\cdot\, \mathbf{cd} = (\mathbf{b} \cdot \mathbf{c})(\mathbf{a} \cdot \mathbf{d})$$

Second rank tensor topics

Transpose

$$\mathbf{A}^T = a^{ji}\mathbf{e}_i\mathbf{e}_j = a_{ji}\mathbf{e}^i\mathbf{e}^j = a^{j}_{\cdot i}\mathbf{e}^i\mathbf{e}_j = a_j^{\cdot i}\mathbf{e}_i\mathbf{e}^j$$

$$\mathbf{A}\cdot\mathbf{x} = \mathbf{x}\cdot\mathbf{A}^T$$

$$(\mathbf{A}^T)^T = \mathbf{A}$$

$$(\mathbf{A}\cdot\mathbf{B})^T = \mathbf{B}^T\cdot\mathbf{A}^T$$

$$\mathbf{a}\cdot\mathbf{C}^T\cdot\mathbf{b} = \mathbf{b}\cdot\mathbf{C}\cdot\mathbf{a}$$

$$\det\mathbf{A}^{-1} = (\det\mathbf{A})^{-1} \qquad (\mathbf{A}\cdot\mathbf{B})^{-1} = \mathbf{B}^{-1}\cdot\mathbf{A}^{-1}$$
$$(\mathbf{A}^T)^{-1} = (\mathbf{A}^{-1})^T \qquad (\mathbf{A}^{-1})^{-1} = \mathbf{A}$$

Tensors raised to powers

$$\mathbf{A}^0 = \mathbf{A} \qquad \mathbf{A}^n = \mathbf{A}\cdot\mathbf{A}^{n-1} \text{ for } n = 1,2,3,\ldots$$

$$\mathbf{A}^{-n} = \mathbf{A}^{-n+1}\cdot\mathbf{A}^{-1} \text{ for } n = 2,3,4,\ldots$$

$$e^{\mathbf{A}} = \mathbf{E} + \frac{\mathbf{A}}{1!} + \frac{\mathbf{A}^2}{2!} + \frac{\mathbf{A}^3}{3!} + \cdots$$

Symmetric and antisymmetric tensors

$$\begin{aligned} &\text{symmetric:} && \mathbf{A} = \mathbf{A}^T; \quad \forall \mathbf{x},\ \mathbf{A} \cdot \mathbf{x} = \mathbf{x} \cdot \mathbf{A} \\ &\text{antisymmetric:} && \mathbf{A} = -\mathbf{A}^T \end{aligned}$$

$$\mathbf{A} = \frac{1}{2}\left(\mathbf{A} + \mathbf{A}^T\right) + \frac{1}{2}\left(\mathbf{A} - \mathbf{A}^T\right)$$

Eigenpair

$$\mathbf{A} \cdot \mathbf{x} = \lambda \mathbf{x}$$

Viète formulas for invariants

$$\begin{aligned} I_1(\mathbf{A}) &= \lambda_1 + \lambda_2 + \lambda_3 && = \operatorname{trace} \mathbf{A} \\ I_2(\mathbf{A}) &= \lambda_1\lambda_2 + \lambda_1\lambda_3 + \lambda_2\lambda_3 \\ I_3(\mathbf{A}) &= \lambda_1\lambda_2\lambda_3 && = \det \mathbf{A} \end{aligned}$$

Orthogonal tensor

$$\mathbf{Q} \cdot \mathbf{Q}^T = \mathbf{Q}^T \cdot \mathbf{Q} = \mathbf{E}$$

Polar decompositions

$$\begin{aligned} &\mathbf{A} = \mathbf{S} \cdot \mathbf{Q} && \mathbf{Q} \text{ orthogonal} \\ &\mathbf{A} = \mathbf{Q} \cdot \mathbf{S}' && \mathbf{S}, \mathbf{S}' \text{ positive definite and symmetric} \end{aligned}$$

Chapter 4

Vector fields

Some rules for differentiating vector functions

$$\frac{d(\mathbf{e}_1(t)+\mathbf{e}_1(t))}{dt} = \frac{d\mathbf{e}_1(t)}{dt} + \frac{d\mathbf{e}_2(t)}{dt}$$
$$\frac{d(c\mathbf{e}_1(t))}{dt} = c\frac{d\mathbf{e}_1(t)}{dt}$$
$$\frac{d(f(t)\mathbf{e}(t))}{dt} = \frac{df(t)}{dt}\mathbf{e}(t) + f(t)\frac{d\mathbf{e}(t)}{dt}$$

$$\frac{d}{dt}(\mathbf{e}_1(t)\cdot\mathbf{e}_2(t)) = \mathbf{e}_1'(t)\cdot\mathbf{e}_2(t) + \mathbf{e}_1(t)\cdot\mathbf{e}_2'(t)$$
$$\frac{d}{dt}(\mathbf{e}_1(t)\times\mathbf{e}_2(t)) = \mathbf{e}_1'(t)\times\mathbf{e}_2(t) + \mathbf{e}_1(t)\times\mathbf{e}_2'(t)$$

and

$$\frac{d}{dt}[\mathbf{e}_1(t)\,\mathbf{e}_2(t)\,\mathbf{e}_3(t)] = [\mathbf{e}_1'(t)\,\mathbf{e}_2(t)\,\mathbf{e}_3(t)] + [\mathbf{e}_1(t)\,\mathbf{e}_2'(t)\,\mathbf{e}_3(t)] + [\mathbf{e}_1(t)\,\mathbf{e}_2(t)\,\mathbf{e}_3'(t)]$$

where

$$[\mathbf{e}_1(t)\,\mathbf{e}_2(t)\,\mathbf{e}_3(t)] = (\mathbf{e}_1(t)\times\mathbf{e}_2(t))\cdot\mathbf{e}_3(t)$$

Tangent vectors to coordinate lines

$$\mathbf{r}_i = \frac{\partial\mathbf{r}}{\partial q^i} \qquad i = 1,2,3$$

Jacobian

$$\sqrt{g} = \mathbf{r}_1\cdot(\mathbf{r}_2\times\mathbf{r}_3) = \left|\frac{\partial x^i}{\partial q^j}\right|$$

Pointwise definition of reciprocal basis

$$\mathbf{r}^i\cdot\mathbf{r}_j = \delta^i_j$$

Definition of metric coefficients

$$g_{ij} = \mathbf{r}_i \cdot \mathbf{r}_j$$
$$g^{ij} = \mathbf{r}^i \cdot \mathbf{r}^j$$
$$g_i^j = \mathbf{r}_i \cdot \mathbf{r}^j = \delta_i^j$$

Transformation laws

$$\mathbf{r}_i = A_i^j \tilde{\mathbf{r}}_j \qquad A_i^j = \frac{\partial \tilde{q}^j}{\partial q^i}$$

$$\tilde{\mathbf{r}}_i = \tilde{A}_i^j \mathbf{r}_j \qquad \tilde{A}_i^j = \frac{\partial q^j}{\partial \tilde{q}^i}$$

$$\tilde{f}^i = A_j^i f^j \qquad \tilde{f}_i = \tilde{A}_i^j f_j \qquad f^i = \tilde{A}_j^i \tilde{f}^j \qquad f_i = A_i^j \tilde{f}_j$$

Differentials and the nabla operator

Metric forms

$$(ds)^2 = d\mathbf{r} \cdot d\mathbf{r} = g_{ij} dq^i dq^j$$

Nabla operator

$$\nabla = \mathbf{r}^i \frac{\partial}{\partial q^i}$$

Gradient of a vector function

$$d\mathbf{f} = d\mathbf{r} \cdot \nabla \mathbf{f} = \nabla \mathbf{f}^T \cdot d\mathbf{r}$$

Divergence of vector

$$\operatorname{div} \mathbf{f} = \nabla \cdot \mathbf{f} = \mathbf{r}^i \cdot \frac{\partial \mathbf{f}}{\partial q^i}$$

Rotation and curl of vector

$$\operatorname{rot} \mathbf{f} = \nabla \times \mathbf{f} = \mathbf{r}^i \times \frac{\partial \mathbf{f}}{\partial q^i} \qquad \boldsymbol{\omega} = \frac{1}{2} \operatorname{rot} \mathbf{f}$$

Divergence and rotation of second-rank tensor

$$\nabla \cdot \mathbf{A} = \mathbf{r}^i \cdot \frac{\partial}{\partial q^i}\mathbf{A} \qquad \nabla \times \mathbf{A} = \mathbf{r}^i \times \frac{\partial}{\partial q^i}\mathbf{A}$$

Differentiation of a vector function

Christoffel coefficients of the second kind

$$\frac{\partial \mathbf{r}_i}{\partial q^j} = \Gamma_{ij}^k \mathbf{r}_k \qquad \frac{\partial \mathbf{r}^j}{\partial q^i} = -\Gamma_{it}^j \mathbf{r}^t$$

$$\Gamma_{ij}^k = \Gamma_{ji}^k$$

Christoffel coefficients of the first kind

$$\frac{1}{2}\left(\frac{\partial g_{it}}{\partial q^j} + \frac{\partial g_{tj}}{\partial q^i} - \frac{\partial g_{ji}}{\partial q^t}\right) = \Gamma_{ijt}$$

$$\Gamma_{ijk} = \Gamma_{jik}$$

Covariant differentiation

$$\frac{\partial \mathbf{f}}{\partial q^i} = \mathbf{r}^k \nabla_i f_k = \mathbf{r}_j \nabla_i f^j$$

$$\nabla_k f_i = \frac{\partial f_i}{\partial q^k} - \Gamma_{ki}^j f_j \qquad \nabla_k f^i = \frac{\partial f^i}{\partial q^k} + \Gamma_{kt}^i f^t$$

$$\nabla \mathbf{f} = \mathbf{r}^i \mathbf{r}^j \nabla_i f_j = \mathbf{r}^i \mathbf{r}_j \nabla_i f^j$$

Covariant differentiation of second-rank tensor

$$\frac{\partial}{\partial q^k}\mathbf{A} = \nabla_k a^{ij} \mathbf{r}_i \mathbf{r}_j = \nabla_k a_{ij} \mathbf{r}^i \mathbf{r}^j = \nabla_k a_i^{\cdot j} \mathbf{r}^i \mathbf{r}_j = \nabla_k a_{\cdot j}^{i} \mathbf{r}_i \mathbf{r}^j$$

$$\nabla_k a^{ij} = \frac{\partial a^{ij}}{\partial q^k} + \Gamma^i_{ks} a^{sj} + \Gamma^j_{ks} a^{is} \qquad \nabla_k a_{ij} = \frac{\partial a_{ij}}{\partial q^k} - \Gamma^s_{ki} a_{sj} - \Gamma^s_{kj} a_{is}$$

$$\nabla_k a_i^{\cdot j} = \frac{\partial a_i^{\cdot j}}{\partial q^k} - \Gamma^s_{ki} a_s^{\cdot j} + \Gamma^j_{ks} a_i^{\cdot s} \qquad \nabla_k a^i_{\cdot j} = \frac{\partial a^i_{\cdot j}}{\partial q^k} + \Gamma^i_{ks} a^s_{\cdot j} - \Gamma^s_{kj} a^i_{\cdot s}$$

Differential operations

$$\nabla \times \mathbf{f} = \mathbf{r}_k \epsilon^{ijk} \frac{\partial f_j}{\partial q^i}$$

$$\Gamma^i_{in} = \frac{1}{\sqrt{g}} \frac{\partial \sqrt{g}}{\partial q^n}$$

$$\nabla \cdot \mathbf{f} = \frac{1}{\sqrt{g}} \frac{\partial}{\partial q^i} \left(\sqrt{g} f^i\right)$$

$$\nabla \times \mathbf{A} = \epsilon^{kin} \mathbf{r}_n \frac{\partial}{\partial q^k} \left(\mathbf{r}^j a_{ij}\right)$$

$$\nabla \cdot \mathbf{A} = \frac{1}{\sqrt{g}} \frac{\partial}{\partial q^i} \left(\sqrt{g} a^{ij} \mathbf{r}_j\right)$$

$$\nabla^2 f \equiv \nabla \cdot \nabla f = g^{ij} \left(\frac{\partial^2 f}{\partial q^i \partial q^j} - \Gamma^k_{ij} \frac{\partial f}{\partial q^k} \right)$$

$$\nabla^2 f = \nabla^j \nabla_j f, \qquad \nabla^j \equiv g^{ij} \nabla_i$$

$$\nabla^2 \mathbf{f} = \mathbf{r}_j \nabla^i \nabla_i f^j$$

$$\nabla \nabla \cdot \mathbf{f} = \mathbf{r}^i \nabla_i \nabla_j f^j$$

$$\nabla \times \nabla \times \mathbf{f} = \nabla \nabla \cdot \mathbf{f} - \nabla^2 \mathbf{f}$$

Orthogonal coordinate systems

Lamé coefficients

$$\begin{aligned} &(H_i)^2 = g_{ii} \\ &\mathbf{r}^i = \mathbf{r}_i/(H_i)^2 \qquad\qquad i = 1,2,3 \\ &\hat{\mathbf{r}}_i = \mathbf{r}_i/H_i \end{aligned}$$

Differentiation in the orthogonal basis

$$\nabla = \frac{\hat{\mathbf{r}}_i}{H_i}\frac{\partial}{\partial q^i}$$

$$\nabla \mathbf{f} = \hat{\mathbf{r}}_i \hat{\mathbf{r}}_j \left(\frac{1}{H_i}\frac{\partial f_j}{\partial q^i} - \frac{f_i}{H_i H_j}\frac{\partial H_i}{\partial q^j} + \delta_{ij}\frac{f_k}{H_k}\frac{1}{H_i}\frac{\partial H_i}{\partial q^k} \right)$$

$$\nabla \cdot \mathbf{f} = \frac{1}{H_1 H_2 H_3}\left(\frac{\partial}{\partial q^1}(H_2 H_3 f_1) + \frac{\partial}{\partial q^2}(H_3 H_1 f_2) + \frac{\partial}{\partial q^3}(H_1 H_2 f_3) \right)$$

$$\nabla \times \mathbf{f} = \frac{1}{2}\frac{\hat{\mathbf{r}}_i \times \hat{\mathbf{r}}_j}{H_i H_j}\left(\frac{\partial}{\partial q^i}(H_j f_j) - \frac{\partial}{\partial q^j}(H_i f_i) \right)$$

$$\begin{aligned} \nabla^2 f = \frac{1}{H_1 H_2 H_3} &\left[\frac{\partial}{\partial q^1}\left(\frac{H_2 H_3}{H_1}\frac{\partial f}{\partial q^1} \right) + \right. \\ &+ \frac{\partial}{\partial q^2}\left(\frac{H_3 H_1}{H_2}\frac{\partial f}{\partial q^2} \right) + \\ &\left. + \frac{\partial}{\partial q^3}\left(\frac{H_1 H_2}{H_3}\frac{\partial f}{\partial q^3} \right) \right] \end{aligned}$$

Integration formulas

Transformation of multiple integral

$$\int_V f(x_1, x_2, x_3)\, dx_1\, dx_2\, dx_3 = \int_V f(q^1, q^2, q^3) J\, dq^1\, dq^2\, dq^3$$

$$J = \sqrt{g} = \left| \frac{\partial x_i}{\partial q^j} \right|$$

Integration by parts

$$\int_V \frac{\partial f}{\partial x_k} g \, dx_1 \, dx_2 \, dx_3 = -\int_V \frac{\partial g}{\partial x_k} f \, dx_1 \, dx_2 \, dx_3 + \int_S f g \, n_k \, dS$$

Miscellaneous results

$$\int_V \nabla f \, dV = \int_S f\mathbf{n} \, dS \qquad \int_V \nabla \cdot \mathbf{f} \, dV = \int_S \mathbf{n} \cdot \mathbf{f} \, dS$$

$$\int_V \nabla \mathbf{f} \, dV = \int_S \mathbf{n}\mathbf{f} \, dS \qquad \int_V \nabla \times \mathbf{f} \, dV = \int_S \mathbf{n} \times \mathbf{f} \, dS$$

$$\int_V \nabla \mathbf{A} \, dV = \int_S \mathbf{n}\mathbf{A} \, dS$$

$$\int_V \nabla \cdot \mathbf{A} \, dV = \int_S \mathbf{n} \cdot \mathbf{A} \, dS$$

$$\int_V \nabla \times \mathbf{A} \, dV = \int_S \mathbf{n} \times \mathbf{A} \, dS$$

$$\oint_\Gamma \mathbf{f} \cdot d\mathbf{r} = \int_S (\mathbf{n} \times \nabla) \cdot \mathbf{f} \, dS$$

$$\oint_\Gamma d\mathbf{r} \cdot \mathbf{A} = \int_S (\mathbf{n} \times \nabla) \cdot \mathbf{A} \, dS$$

$$\oint_\Gamma \mathbf{A} \cdot d\mathbf{r} = \int_S (\mathbf{n} \times \nabla) \cdot \mathbf{A}^T \, dS$$

Chapter 5

Elementary theory of curves

Parameterization

$$\mathbf{r} = \mathbf{r}(t) \qquad \text{or} \qquad \mathbf{r} = \mathbf{r}(s)$$

Length

$$s = \int_a^b |\mathbf{r}'(t)| \, dt$$

Unit tangent

$$\boldsymbol{\tau}(s) = \mathbf{r}'(s)$$

Equation of tangent line

$$\mathbf{r} = \mathbf{r}(t_0) + \lambda\mathbf{r}'(t_0)$$

$$\frac{x - x(t_0)}{x'(t_0)} = \frac{y - y(t_0)}{y'(t_0)} = \frac{z - z(t_0)}{z'(t_0)}$$

Curvature

$$k_1 = |\mathbf{r}''(s)| \qquad k_1^2 = \frac{(\mathbf{r}'(t) \times \mathbf{r}''(t))^2}{(\mathbf{r}'^2(t))^3}$$

Radius of curvature

$$R = 1/k_1$$

Principal normal, binormal

$$\boldsymbol{\nu} = \frac{\mathbf{r}''(s)}{k_1} \qquad \boldsymbol{\beta} = \boldsymbol{\tau} \times \boldsymbol{\nu}$$

Osculating plane

$$[\mathbf{r} - \mathbf{r}(s_0)] \cdot \boldsymbol{\beta}(s_0) = 0$$

$$\begin{vmatrix} x - x(t_0) & y - y(t_0) & z - z(t_0) \\ x'(t_0) & y'(t_0) & z'(t_0) \\ x''(t_0) & y''(t_0) & z''(t_0) \end{vmatrix} = 0$$

Torsion

$$k_2 = -\frac{(\mathbf{r}'(s) \times \mathbf{r}''(s)) \cdot \mathbf{r}'''(s)}{k_1^2} \qquad k_2 = -\frac{(\mathbf{r}'(t) \times \mathbf{r}''(t)) \cdot \mathbf{r}'''(t)}{(\mathbf{r}'(t) \times \mathbf{r}''(t))^2}$$

Serret–Frenet equations

$$\begin{aligned}\boldsymbol{\tau}' &= k_1 \boldsymbol{\nu} \\ \boldsymbol{\nu}' &= -k_1 \boldsymbol{\tau} - k_2 \boldsymbol{\beta} \\ \boldsymbol{\beta}' &= k_2 \boldsymbol{\nu}\end{aligned}$$

Theory of surfaces

Parameterization

$$\mathbf{r} = \mathbf{r}(u^1, u^2)$$

Tangent vectors, unit normal

$$\mathbf{r}_i = \frac{\partial \mathbf{r}}{\partial u^i} \qquad i = 1, 2$$

$$\mathbf{n} = \frac{\mathbf{r}_1 \times \mathbf{r}_2}{|\mathbf{r}_1 \times \mathbf{r}_2|}$$

First fundamental form

$$(ds)^2 = g_{ij} du^i du^j = E(du^1)^2 + 2F du^1 du^2 + G(du^2)^2$$

$$E = \mathbf{r}_1 \cdot \mathbf{r}_1 \qquad F = \mathbf{r}_1 \cdot \mathbf{r}_2 \qquad G = \mathbf{r}_2 \cdot \mathbf{r}_2$$

Orthogonality of curves

$$E du^1 d\tilde{u}^1 + F(du^1 d\tilde{u}^2 + du^2 d\tilde{u}^1) + G du^2 d\tilde{u}^2 = 0$$

Area

$$S = \int_A \sqrt{EG - F^2}\, du^1\, du^2$$

Second fundamental form

$$d^2\mathbf{r} \cdot \mathbf{n} = L(du^1)^2 + 2M du^1 du^2 + N(du^2)^2 = -d\mathbf{r} \cdot d\mathbf{n}$$

$$L = \frac{\partial^2 \mathbf{r}}{(\partial u^1)^2} \cdot \mathbf{n} \qquad M = \frac{\partial^2 \mathbf{r}}{\partial u^1 \partial u^2} \cdot \mathbf{n} \qquad N = \frac{\partial^2 \mathbf{r}}{(\partial u^2)^2} \cdot \mathbf{n}$$

Normal curvature, mean curvature, Gaussian curvature

$$k_0 = k_1 \cos\vartheta \qquad \vartheta = \text{ angle between } \boldsymbol{\nu} \text{ and } \mathbf{n}$$

$$H = \frac{1}{2}(k_{\min} + k_{\max}) = \frac{1}{2}\frac{LG - 2MF + NE}{EG - F^2}$$

$$K = k_{\min} k_{\max} = \frac{LN - M^2}{EG - F^2}$$

Surface given by $z = f(x, y)$ in Cartesian coordinates

Subscripts x, y denote partial derivatives with respect to x, y respectively.

$$E = 1 + {f_x}^2 \qquad F = f_x f_y \qquad G = 1 + {f_y}^2$$

$$EG - F^2 = 1 + {f_x}^2 + {f_y}^2 \qquad S = \int_D \sqrt{1 + {f_x}^2 + {f_y}^2}\, dx\, dy$$

$$\mathbf{n} = \frac{-f_x \mathbf{i}_1 - f_y \mathbf{i}_2 + \mathbf{i}_3}{\sqrt{1 + {f_x}^2 + {f_y}^2}}$$

$$L = \mathbf{r}_{xx} \cdot \mathbf{n} = \frac{f_{xx}}{\sqrt{1 + {f_x}^2 + {f_y}^2}}$$

$$M = \mathbf{r}_{xy} \cdot \mathbf{n} = \frac{f_{xy}}{\sqrt{1 + {f_x}^2 + {f_y}^2}}$$

$$N = \mathbf{r}_{yy} \cdot \mathbf{n} = \frac{f_{yy}}{\sqrt{1 + {f_x}^2 + {f_y}^2}}$$

$$K = \frac{f_{xx} f_{yy} - {f_{xy}}^2}{\left(1 + {f_x}^2 + {f_y}^2\right)^2}$$

Surface of revolution about z-axis

$$x = \phi(u) \qquad z = \psi(u)$$

$$(ds)^2 = \left(\phi'^2 + \psi'^2\right) du^2 + \phi^2 \, dv^2$$

$$-d\mathbf{n} \cdot d\mathbf{r} = \frac{\psi''\phi' - \phi''\psi'}{\sqrt{\phi'^2 + \psi'^2}} \, du^2 + \frac{\psi'\phi}{\sqrt{\phi'^2 + \psi'^2}} \, dv^2$$

Appendix B

Hints and Answers

Chapter 1

Exercise 1.1 The given equation states that $\mathbf{a}, \mathbf{b}, \mathbf{c}$ are not coplanar; $\mathbf{a}$ has a nonzero component perpendicular to the plane of $\mathbf{b}$ and $\mathbf{c}$, hence $\mathbf{a}$ cannot be in this plane.

Chapter 2

Exercise 2.1 Suppose that $\mathbf{e}^i$ and $\mathbf{f}^i$ are two vectors such that $x^i = \mathbf{x} \cdot \mathbf{e}^i$ and $x^i = \mathbf{x} \cdot \mathbf{f}^i$ for all $\mathbf{x}$. Then $\mathbf{x} \cdot (\mathbf{e}^i - \mathbf{f}^i) = 0$ for all $\mathbf{x}$, from which it follows that $\mathbf{e}^i - \mathbf{f}^i = 0$. This shows that $\mathbf{f}^i = \mathbf{e}^i$.

Exercise 2.2 We must show that the equation $\alpha_1 \mathbf{e}^1 + \alpha_2 \mathbf{e}^2 + \alpha_3 \mathbf{e}^3 = \mathbf{0}$ implies $\alpha_1 = \alpha_2 = \alpha_3 = 0$. To get $\alpha_1 = 0$, for example, we simply dot-multiply the equation by $\mathbf{e}_1$ and use $\mathbf{e}_i \cdot \mathbf{e}^j = \delta_i^j$.

Exercise 2.3 We have

$$\begin{aligned}
\mathbf{e}^2 \times \mathbf{e}^3 &= \frac{1}{V^2}(\mathbf{e}_3 \times \mathbf{e}_1) \times (\mathbf{e}_1 \times \mathbf{e}_2) \\
&= \frac{1}{V^2}\{[\mathbf{e}_2 \cdot (\mathbf{e}_3 \times \mathbf{e}_1)]\mathbf{e}_1 - [\mathbf{e}_1 \cdot (\mathbf{e}_3 \times \mathbf{e}_1)]\mathbf{e}_2\} \\
&= \frac{1}{V^2}\mathbf{e}_1[\mathbf{e}_2 \cdot (\mathbf{e}_3 \times \mathbf{e}_1)]
\end{aligned}$$

by the vector triple product identity, hence

$$V' = \frac{1}{V^2}\mathbf{e}^1 \cdot \mathbf{e}_1[\mathbf{e}_2 \cdot (\mathbf{e}_3 \times \mathbf{e}_1)] = \frac{1}{V^2}[\mathbf{e}_1 \cdot (\mathbf{e}_2 \times \mathbf{e}_3)] = \frac{1}{V}.$$

Exercise 2.4 (b) First use common properties of determinants to establish the identity

$$[\mathbf{a}\cdot(\mathbf{b}\times\mathbf{c})][\mathbf{u}\cdot(\mathbf{v}\times\mathbf{w})] = \begin{vmatrix} \mathbf{a}\cdot\mathbf{u} & \mathbf{a}\cdot\mathbf{v} & \mathbf{a}\cdot\mathbf{w} \\ \mathbf{b}\cdot\mathbf{u} & \mathbf{b}\cdot\mathbf{v} & \mathbf{b}\cdot\mathbf{w} \\ \mathbf{c}\cdot\mathbf{u} & \mathbf{c}\cdot\mathbf{v} & \mathbf{c}\cdot\mathbf{w} \end{vmatrix}. \tag{*}$$

Write

$$[\mathbf{a}\cdot(\mathbf{b}\times\mathbf{c})][\mathbf{u}\cdot(\mathbf{v}\times\mathbf{w})] = \begin{vmatrix} a_1 & a_2 & a_3 \\ b_1 & b_2 & b_3 \\ c_1 & c_2 & c_3 \end{vmatrix} \begin{vmatrix} u_1 & u_2 & u_3 \\ v_1 & v_2 & v_3 \\ w_1 & w_2 & w_3 \end{vmatrix}$$

in a Cartesian frame. Then

$$[\mathbf{a}\cdot(\mathbf{b}\times\mathbf{c})][\mathbf{u}\cdot(\mathbf{v}\times\mathbf{w})] = \begin{vmatrix} a_1 & a_2 & a_3 \\ b_1 & b_2 & b_3 \\ c_1 & c_2 & c_3 \end{vmatrix} \begin{vmatrix} u_1 & v_1 & w_1 \\ u_2 & v_2 & w_2 \\ u_3 & v_3 & w_3 \end{vmatrix}$$

$$= \begin{vmatrix} a_1u_1 + a_2u_2 + a_3u_3 & a_1v_1 + a_2v_2 + a_3v_3 & a_1w_1 + a_2w_2 + a_3w_3 \\ b_1u_1 + b_2u_2 + b_3u_3 & b_1v_1 + b_2v_2 + b_3v_3 & b_1w_1 + b_2w_2 + b_3w_3 \\ c_1u_1 + c_2u_2 + c_3u_3 & c_1v_1 + c_2v_2 + c_3v_3 & c_1w_1 + c_2w_2 + c_3w_3 \end{vmatrix}.$$

(Note that our intermediate steps occurred in a Cartesian frame, but the final result is still frame independent.) Finally, write

$$V^2 = [\mathbf{e}_1\cdot(\mathbf{e}_2\times\mathbf{e}_3)]^2 = \begin{vmatrix} \mathbf{e}_1\cdot\mathbf{e}_1 & \mathbf{e}_1\cdot\mathbf{e}_2 & \mathbf{e}_1\cdot\mathbf{e}_3 \\ \mathbf{e}_2\cdot\mathbf{e}_1 & \mathbf{e}_2\cdot\mathbf{e}_2 & \mathbf{e}_2\cdot\mathbf{e}_3 \\ \mathbf{e}_3\cdot\mathbf{e}_1 & \mathbf{e}_3\cdot\mathbf{e}_2 & \mathbf{e}_3\cdot\mathbf{e}_3 \end{vmatrix} = g.$$

Exercise 2.5

(a) $|\mathbf{x}|^2 = x_k x^k = g_{ij}x^i x^j = g^{mn}x_m x_n$.
(b) $\mathbf{x}\cdot\mathbf{y} = x^k y_k = x_k y^k = g_{ij}x^i y^j = g^{mn}x_m y_n$.

Exercise 2.6 The transformation matrix is of the form

$$\begin{pmatrix} \cos\theta & \sin\theta & 0 \\ -\sin\theta & \cos\theta & 0 \\ 0 & 0 & 1 \end{pmatrix}$$

where θ is the angle of rotation.

Exercise 2.7 Equate two different expressions for $\mathbf{x}$ as is done in the text.

Exercise 2.8

$$\epsilon^{ijk} = \begin{cases} +V', & (i,j,k) \text{ an even permutation of } (1,2,3), \\ -V', & (i,j,k) \text{ an odd permutation of } (1,2,3), \\ 0, & \text{two or more indices equal.} \end{cases}$$

(Recall that $V' = 1/V$.) Then use the determinantal identity established in Exercise 2.4: put $\mathbf{a} = \mathbf{e}_i$, $\mathbf{b} = \mathbf{e}_j$, $\mathbf{c} = \mathbf{e}_k$, $\mathbf{u} = \mathbf{e}^p$, $\mathbf{v} = \mathbf{e}^q$, $\mathbf{w} = \mathbf{e}^r$. To prove the vector triple product identity we write

$$\begin{aligned} \mathbf{a} \times (\mathbf{b} \times \mathbf{c}) &= \mathbf{e}^i \epsilon_{ijk} a^j \epsilon^{kpq} b_p c_q = \mathbf{e}^i (\delta_i^p \delta_j^q - \delta_i^q \delta_j^p) a^j b_p c_q \\ &= \mathbf{e}^i b_i (a^q c_q) - \mathbf{e}^i c_i (a^p b_p). \end{aligned}$$

Exercise 2.9 Using the scalar triple product identity and (2.13) we have

$$\begin{aligned} (\mathbf{a} \times \mathbf{b}) \cdot (\mathbf{c} \times \mathbf{d}) &= \mathbf{a} \cdot [\mathbf{b} \times (\mathbf{c} \times \mathbf{d})] \\ &= \mathbf{a} \cdot [(\mathbf{b} \cdot \mathbf{d})\mathbf{c} - (\mathbf{b} \cdot \mathbf{c})\mathbf{d}], \end{aligned}$$

and the result follows.

Chapter 3

Exercise 3.1 (b)

$$\begin{pmatrix} 1 & 0 & 0 \\ 0 & 0 & 0 \\ 0 & 0 & 0 \end{pmatrix}, \qquad \begin{pmatrix} 0 & 0 & 0 \\ 0 & 1 & 0 \\ 0 & 0 & 0 \end{pmatrix}, \qquad \begin{pmatrix} 0 & 0 & 0 \\ 0 & 0 & 0 \\ 1 & 0 & 0 \end{pmatrix}.$$

Exercise 3.2 $\mathbf{E}$ can be written in any of the forms

$$\mathbf{E} = e^{ij} \mathbf{e}_i \mathbf{e}_j = e_{ij} \mathbf{e}^i \mathbf{e}^j = e^i_{\cdot j} \mathbf{e}_i \mathbf{e}^j = e_i^{\cdot j} \mathbf{e}^i \mathbf{e}_j.$$

Let us choose the first form as an example. Equation (3.4) with $\mathbf{x} = \mathbf{e}^k$ implies $e^{ij} \mathbf{e}_i \mathbf{e}_j \cdot \mathbf{e}^k = \mathbf{e}^k$. Pre-dotting this with $\mathbf{e}^m$ yields $e^{mk} = g^{mk}$.

Exercise 3.3 $(\mathbf{A} \cdot \mathbf{B}) \cdot (\mathbf{B}^{-1} \cdot \mathbf{A}^{-1}) = \mathbf{A} \cdot (\mathbf{B} \cdot \mathbf{B}^{-1}) \cdot \mathbf{A}^{-1} = \mathbf{A} \cdot \mathbf{E} \cdot \mathbf{A}^{-1} = \mathbf{A} \cdot \mathbf{A}^{-1} = \mathbf{E}$.

Exercise 3.4 (a) $\tilde{B} = A^T B A$. (b) $\tilde{B} = A B A^T$.

Exercise 3.5 We are given $\mathbf{e}_i = A_i^j \tilde{\mathbf{e}}_j$. Now it is necessary to understand what the needed tensor does. We denote it by $\mathbf{A}$. It is the map $\mathbf{y} = \mathbf{A} \cdot \mathbf{x}$ that must take $\mathbf{x} = \mathbf{e}_i$ into $\mathbf{y} = A_i^j \tilde{\mathbf{e}}_j$ for each $i = 1, 2, 3$. Let us write out these equations:

$$\mathbf{A} \cdot \mathbf{e}_i = A_i^j \tilde{\mathbf{e}}_j.$$

The decisive step is to suppose that $\mathbf{A}$ should be written in a "mixed basis" as $\mathbf{A} = a^{m}_{\cdot n} \tilde{\mathbf{e}}_m \mathbf{e}^n$. We have

$$a^{m}_{\cdot n} \tilde{\mathbf{e}}_m \mathbf{e}^n \cdot \mathbf{e}_i = A_i^j \tilde{\mathbf{e}}_j$$

and it follows that

$$a^{m}_{\cdot i} \tilde{\mathbf{e}}_m = A_i^j \tilde{\mathbf{e}}_j.$$

Thus (because the indices are dummy)

$$a^{j}_{\cdot i} = A_i^j$$

and so the tensor is $\mathbf{A} = a^{m}_{\cdot n} \tilde{\mathbf{e}}_m \mathbf{e}^n$ where $a^{m}_{\cdot n} = A_n^m$.

Exercise 3.7

(a) $a^{ij} \mathbf{e}_i \mathbf{e}_j \cdot\cdot \mathbf{e}^k \mathbf{e}_k = a^{ij} \delta_j^k g_{ik} = a_k^{\cdot k}$.

(b) $a_{ij} \mathbf{e}^i \mathbf{e}^j \cdot\cdot b^{kn} \mathbf{e}_k \mathbf{e}_n = a_{ij} b^{kn} \delta_k^j \delta_n^i = a_{ik} b^{ki}$. Having obtained this we may use the metric tensor to raise and lower indices and thereby obtain other forms; for example, $a_{ik} b^{ki} = a^{p}_{\cdot k} g_{pi} b^{ki} = a^{p}_{\cdot k} b^{k}_{\cdot p}$.

Exercise 3.8

(a) Write $\mathbf{A} = a^{ij} \mathbf{e}_i \mathbf{e}_j$, $\mathbf{B} = b^{kn} \mathbf{e}_k \mathbf{e}_n$, $\mathbf{A}^T = a^{ji} \mathbf{e}_i \mathbf{e}_j$, $\mathbf{B}^T = b^{nk} \mathbf{e}_k \mathbf{e}_n$, and show that both sides can be written as $a^{jp} b_p^{\cdot k} \mathbf{e}_k \mathbf{e}_j$.

(b) $\mathbf{b} \cdot \mathbf{C} \cdot \mathbf{a} = \mathbf{b} \cdot (\mathbf{C} \cdot \mathbf{a}) = \mathbf{b} \cdot (\mathbf{a} \cdot \mathbf{C}^T) = (\mathbf{a} \cdot \mathbf{C}^T) \cdot \mathbf{b} = \mathbf{a} \cdot \mathbf{C}^T \cdot \mathbf{b}$.

(c) See the reference by Lurie.

(d) The relation $(\mathbf{A}^T)^{-1} = (\mathbf{A}^{-1})^T$ follows from the two relations

$$\mathbf{A}^T \cdot (\mathbf{A}^{-1})^T = (\mathbf{A}^{-1} \cdot \mathbf{A})^T = \mathbf{E}^T = \mathbf{E},$$
$$(\mathbf{A}^{-1})^T \cdot \mathbf{A}^T = (\mathbf{A} \cdot \mathbf{A}^{-1})^T = \mathbf{E}^T = \mathbf{E}.$$

Exercise 3.9

$$\begin{pmatrix} a^{11} & a^{12} & a^{13} \\ a^{12} & a^{22} & a^{23} \\ a^{13} & a^{23} & a^{33} \end{pmatrix}, \qquad \begin{pmatrix} 0 & a^{12} & a^{13} \\ -a^{12} & 0 & a^{23} \\ -a^{13} & -a^{23} & 0 \end{pmatrix}.$$

Exercise 3.10 With $\mathbf{A} = a^{ij}\mathbf{e}_i\mathbf{e}_j$ and $\mathbf{B} = b^{kn}\mathbf{e}_k\mathbf{e}_n$ we have

$$\mathbf{A} \cdot\cdot \mathbf{B} = a^{ij}b^{kn}g_{jk}g_{in}. \qquad (*)$$

Since $\mathbf{A}$ is symmetric and $\mathbf{B}$ is antisymmetric, we may also write $\mathbf{A} \cdot\cdot \mathbf{B} = -a^{ji}b^{nk}g_{jk}g_{in}$; let us swap i with j and k with n in this expression to get

$$\mathbf{A} \cdot\cdot \mathbf{B} = -a^{ij}b^{kn}g_{in}g_{jk}. \qquad (**)$$

Adding (*) and (**) we obtain $2(\mathbf{A} \cdot\cdot \mathbf{B}) = 0$.

Exercise 3.11

$$\begin{aligned}\mathbf{x} \cdot \left[\frac{1}{2}(\mathbf{A} + \mathbf{A}^T)\right] \cdot \mathbf{x} &= \frac{1}{2}\left[\mathbf{x} \cdot \mathbf{A} \cdot \mathbf{x} + \mathbf{x} \cdot \mathbf{A}^T \cdot \mathbf{x}\right] \\ &= \frac{1}{2}\left[\mathbf{x} \cdot \mathbf{A} \cdot \mathbf{x} + \mathbf{x} \cdot \mathbf{A} \cdot \mathbf{x}\right] \\ &= \mathbf{x} \cdot \mathbf{A} \cdot \mathbf{x}.\end{aligned}$$

Exercise 3.12 They are (1) $\mathbf{x} = \mathbf{a}$ with $\lambda = \mathbf{b} \cdot \mathbf{a}$, and (2) any $\mathbf{x}$ that is perpendicular to $\mathbf{b}$, with $\lambda = 0$.

Exercise 3.13 The characteristic equation $-\lambda^3 + 2\lambda^2 + \lambda = 0$ has solutions $\lambda_1 = 0$, $\lambda_2 = 1 + \sqrt{2}$, $\lambda_3 = 1 - \sqrt{2}$. The Viète formulas give

$$\begin{aligned}I_1(\mathbf{A}) &= \lambda_1 + \lambda_2 + \lambda_3 = 2, \\ I_2(\mathbf{A}) &= \lambda_1\lambda_2 + \lambda_1\lambda_3 + \lambda_2\lambda_3 = -1, \\ I_3(\mathbf{A}) &= \lambda_1\lambda_2\lambda_3 = 0.\end{aligned}$$

Exercise 3.14 We prove this for two eigenvectors; the reader may readily generalize to any number of eigenvectors. Let $\mathbf{A} \cdot \mathbf{x}_1 = \lambda_1\mathbf{x}_1$ and $\mathbf{A} \cdot \mathbf{x}_2 = \lambda_2\mathbf{x}_2$ where $\lambda_2 \neq \lambda_1$. Now suppose

$$\alpha_1\mathbf{x}_1 + \alpha_2\mathbf{x}_2 = \mathbf{0}. \qquad (*)$$

Let us operate on both sides of (*) with the tensor $\mathbf{A} - \lambda_2\mathbf{I}$. After simplification we obtain

$$\alpha_1(\lambda_1 - \lambda_2)\mathbf{x}_1 = \mathbf{0}.$$

Because an eigenvector is nonzero by definition we have $\alpha_1 = 0$. Putting this back into (*) we also have $\alpha_2 = 0$.

Exercise 3.15 Assume $\mathbf{x}_1, \mathbf{x}_2, \mathbf{x}_3$ are eigenvectors of $\mathbf{A}$ corresponding to the distinct eigenvalues $\lambda_1, \lambda_2, \lambda_3$, respectively. Let us show that any eigenvector corresponding to λ_1 must be a scalar multiple of $\mathbf{x}_1$. So suppose that

$$\mathbf{A} \cdot \mathbf{x} = \lambda_1 \mathbf{x}. \qquad (**)$$

By the linear independence of $\mathbf{x}_1, \mathbf{x}_2, \mathbf{x}_3$ these vectors form a basis for three-dimensional space and we can express

$$\mathbf{x} = c_1 \mathbf{x}_1 + c_2 \mathbf{x}_2 + c_3 \mathbf{x}_3.$$

Applying $\mathbf{A}$ to both sides we have, by (**),

$$\lambda_1 \mathbf{x} = \lambda_1 c_1 \mathbf{x}_1 + \lambda_1 c_2 \mathbf{x}_2 + \lambda_1 c_3 \mathbf{x}_3 = c_1 \lambda_1 \mathbf{x}_1 + c_2 \lambda_2 \mathbf{x}_2 + c_3 \lambda_3 \mathbf{x}_3.$$

This simplifies to

$$(\lambda_1 - \lambda_2) c_2 \mathbf{x}_2 + (\lambda_1 - \lambda_3) c_3 \mathbf{x}_3 = \mathbf{0}.$$

By linear independence $c_2 = c_3 = 0$. Hence $\mathbf{x} = c_1 \mathbf{x}_1$.

Exercise 3.16 The characteristic equation of $\mathbf{A}$ is

$$\begin{vmatrix} 1-\lambda & 0 & 0 \\ 1 & 1-\lambda & 0 \\ 0 & 1 & -\lambda \end{vmatrix} = 0$$

or $\lambda^3 - 2\lambda^2 + \lambda = 0$. Thus $\mathbf{A}^3 = 2\mathbf{A}^2 - \mathbf{A}$.

Exercise 3.17

(a) $\mathbf{Q} \cdot \mathbf{Q}^T = \mathbf{Q}^T \cdot \mathbf{Q} = \mathbf{i}_1 \mathbf{i}_1 + \mathbf{i}_2 \mathbf{i}_2 + \mathbf{i}_3 \mathbf{i}_3 = \mathbf{E}$. Since we can write $\mathbf{Q} = -\mathbf{i}_1 \mathbf{i}^1 + \mathbf{i}_2 \mathbf{i}^2 + \mathbf{i}_3 \mathbf{i}^3$, the matrix of $\mathbf{Q}$ in mixed components is

$$\begin{pmatrix} -1 & 0 & 0 \\ 0 & 1 & 0 \\ 0 & 0 & 1 \end{pmatrix}.$$

Hence $\det \mathbf{Q} = -1$.

(b) The defining equation $\mathbf{Q} \cdot \mathbf{Q}^T = \mathbf{E}$ implies

$$q_{ij} \mathbf{e}^i \mathbf{e}^j \cdot q^{km} \mathbf{e}_m \mathbf{e}_k = q_{ij} q^{kj} \mathbf{e}^i \mathbf{e}_k = \delta_i^k \mathbf{e}^i \mathbf{e}_k.$$

(c) If n is a positive integer,

$$\mathbf{Q}^n \cdot (\mathbf{Q}^n)^T = \mathbf{Q}^n \cdot (\mathbf{Q}^T)^n = (\mathbf{Q} \cdot \mathbf{Q}^T)^n = \mathbf{E}^n = \mathbf{E}.$$

Exercise 3.18 Pre-dot $\mathbf{A} \cdot \mathbf{x} = \lambda \mathbf{x}$ with $\mathbf{x}$ to get

$$\lambda = \frac{\mathbf{x} \cdot (\mathbf{A} \cdot \mathbf{x})}{|\mathbf{x}|^2}.$$

Clearly $\lambda > 0$ under the stated condition.

Exercise 3.19

(a) $\boldsymbol{\mathcal{E}} \cdots \mathbf{zyx} = \epsilon_{ijk}(\mathbf{e}^k \cdot \mathbf{z})(\mathbf{e}^j \cdot \mathbf{y})(\mathbf{e}^i \cdot \mathbf{x}) = x^i(\epsilon_{ijk} y^j z^k)$.
(b) $\boldsymbol{\mathcal{E}} \cdot\cdot\, \mathbf{xy} = \mathbf{e}^i \epsilon_{ijk}(\mathbf{e}^k \cdot \mathbf{x})(\mathbf{e}^j \cdot \mathbf{y}) = \mathbf{e}^i \epsilon_{ijk} y^j x^k$.
(c) $\boldsymbol{\mathcal{E}} \cdot \mathbf{x} = \epsilon_{ijk} \mathbf{e}^i \mathbf{e}^j x^k$ so that $(\boldsymbol{\mathcal{E}} \cdot \mathbf{x}) \cdot \mathbf{y} = \epsilon_{ijk} \mathbf{e}^i y^j x^k = \mathbf{y} \times \mathbf{x}$.

Chapter 4

Exercise 4.1

(a) Consider the derivative of the dot product. Since we are given the vectors explicitly, an easy way is to dot the vectors first:

$$\mathbf{e}_1(t) \cdot \mathbf{e}_2(t) = e^{-t}(1 - \sin^2 t) = e^{-t} \cos^2 t,$$

hence

$$[\mathbf{e}_1(t) \cdot \mathbf{e}_2(t)]' = -e^{-t} \cos t (2 \sin t + \cos t).$$

However, it is easily checked that the product-rule identity given in the text yields the same result.
(b) $[\mathbf{e}(t) \times \mathbf{e}'(t)]' = \mathbf{e}'(t) \times \mathbf{e}'(t) + \mathbf{e}(t) \times \mathbf{e}''(t) = \mathbf{e}(t) \times \mathbf{e}''(t)$.

Exercise 4.3 Use the product rule to take the indicated second derivative of $\mathbf{r} = \hat{\boldsymbol{\rho}}\rho$, noting that

$$\frac{d\hat{\boldsymbol{\rho}}}{dt} = \frac{d}{dt}(\hat{\mathbf{x}} \cos\phi + \hat{\mathbf{y}} \sin\phi) = (-\hat{\mathbf{x}} \sin\phi + \hat{\mathbf{y}} \cos\phi)\frac{d\phi}{dt} = \hat{\boldsymbol{\phi}}\frac{d\phi}{dt}$$

and, similarly,

$$\frac{d\hat{\boldsymbol{\phi}}}{dt} = -\hat{\boldsymbol{\rho}}\frac{d\phi}{dt}.$$

Exercise 4.4 $v^1 = v_r$, $v^2 = v_\theta/r$, $v^3 = v_\phi/r\sin\theta$.

Exercise 4.5

(a) Fix $v = v_0$ and eliminate u from the transformation equations to get

$$y = x\tan\alpha + v_0(\sin\beta - \cos\beta\tan\alpha);$$

thus the $v = v_0$ coordinate curve is a straight line in the xy-plane making an angle α with the x-axis. Similarly, the $u = u_0$ line makes an angle β with the x-axis.

(b) $\mathbf{r} = \hat{\mathbf{x}}x + \hat{\mathbf{y}}y = \hat{\mathbf{x}}(u\cos\alpha + v\cos\beta) + \hat{\mathbf{y}}(u\sin\alpha + v\sin\beta)$ gives

$$\mathbf{r}_1 = \frac{\partial\mathbf{r}}{\partial u} = \hat{\mathbf{x}}\cos\alpha + \hat{\mathbf{y}}\sin\alpha, \qquad \mathbf{r}_2 = \frac{\partial\mathbf{r}}{\partial v} = \hat{\mathbf{x}}\cos\beta + \hat{\mathbf{y}}\sin\beta.$$

Note that these are both unit vectors. To find $\mathbf{r}^1$ we write $\mathbf{r}^1 = \hat{\mathbf{x}}a + \hat{\mathbf{y}}b$ and solve the equations

$$\mathbf{r}^1\cdot\mathbf{r}_1 = 1, \qquad \mathbf{r}^1\cdot\mathbf{r}_2 = 0,$$

simultaneously to get a, b. By this method we find that

$$\mathbf{r}^1 = \hat{\mathbf{x}}\frac{\sin\beta}{\sin(\beta-\alpha)} - \hat{\mathbf{y}}\frac{\cos\beta}{\sin(\beta-\alpha)},$$
$$\mathbf{r}^2 = -\hat{\mathbf{x}}\frac{\sin\alpha}{\sin(\beta-\alpha)} + \hat{\mathbf{y}}\frac{\cos\alpha}{\sin(\beta-\alpha)}.$$

These are not unit vectors (the coordinate system is non-orthogonal) except in such trivial cases as $\beta - \alpha = \pi/2$. The metric coefficients are

$$g_{11} = 1 = g_{22}, \qquad g_{12} = g_{21} = \cos(\beta-\alpha),$$

and

$$g^{11} = \frac{1}{\sin^2(\beta-\alpha)} = g^{22}, \qquad g^{12} = g^{21} = -\frac{\cos(\beta-\alpha)}{\sin^2(\beta-\alpha)}.$$

(c) Write $\mathbf{z} = z^1\mathbf{r}_1 + z^2\mathbf{r}_2 = z_1\mathbf{r}^1 + z_2\mathbf{r}^2$. Substitute for the $\mathbf{r}_i$ and $\mathbf{r}^i$ in terms of $\hat{\mathbf{x}}$ and $\hat{\mathbf{y}}$ and then equate coefficients of $\hat{\mathbf{x}}, \hat{\mathbf{y}}$ to get simultaneous equations for the z_i in terms of the z^i. The answers are

$$z_1 = z^1 + z^2\cos(\beta-\alpha), \qquad z_2 = z^1\cos(\beta-\alpha) + z^2.$$

Exercise 4.6 $(ds)^2 = (du)^2 + 2\,du\,dv\,\cos(\beta - \alpha) + (dv)^2$.

Exercise 4.7

(a) The total differential of φ is

$$d\varphi = \frac{\partial\varphi}{\partial q^i}\,dq^i.$$

If we start at a point on the given surface and move in such a way that we stay on the surface, then $d\varphi = 0$. In the text it is shown that $dq^i = \mathbf{r}^i \cdot d\mathbf{r}$, so if we move along $d\mathbf{r}$ satisfying

$$\frac{\partial\varphi}{\partial q^i}\mathbf{r}^i \cdot d\mathbf{r} = 0$$

then we will stay on the surface. It is clear that the vector

$$\nabla\varphi = \frac{\partial\varphi}{\partial q^i}\mathbf{r}^i$$

is normal to the surface, since it is perpendicular to every tangent direction. We have

$$\begin{aligned} |\nabla\varphi| &= \sqrt{\nabla\varphi \cdot \nabla\varphi} \\ &= \sqrt{\frac{\partial\varphi}{\partial q^m}\mathbf{r}^m \cdot \frac{\partial\varphi}{\partial q^n}\mathbf{r}^n} \\ &= \sqrt{g^{mn}\frac{\partial\varphi}{\partial q^m}\frac{\partial\varphi}{\partial q^n}}, \end{aligned}$$

hence

$$\mathbf{n} = \frac{\nabla\varphi}{|\nabla\varphi|} = \frac{\dfrac{\partial\varphi}{\partial q^i}\mathbf{r}^i}{\sqrt{g^{mn}\dfrac{\partial\varphi}{\partial q^m}\dfrac{\partial\varphi}{\partial q^n}}} = \frac{g^{ij}\dfrac{\partial\varphi}{\partial q^i}}{\sqrt{g^{mn}\dfrac{\partial\varphi}{\partial q^m}\dfrac{\partial\varphi}{\partial q^n}}}\mathbf{r}_j.$$

(b) Use the result of part (a).
(c) Use the result of part (b).
(d) Use the result of part (b).

Exercise 4.8 For cylindrical we have

$$\begin{aligned}\nabla f &= \mathbf{r}^i \frac{\partial f}{\partial q^i} \\ &= \mathbf{r}^1 \frac{\partial f}{\partial \rho} + \mathbf{r}^2 \frac{\partial f}{\partial \phi} + \mathbf{r}^3 \frac{\partial f}{\partial z} \\ &= \mathbf{r}_1 \frac{\partial f}{\partial \rho} + \frac{\mathbf{r}_2}{\rho^2} \frac{\partial f}{\partial \phi} + \mathbf{r}_3 \frac{\partial f}{\partial z} \\ &= \hat{\boldsymbol{\rho}} \frac{\partial f}{\partial \rho} + \frac{\hat{\boldsymbol{\phi}}}{\rho} \frac{\partial f}{\partial \phi} + \hat{\mathbf{z}} \frac{\partial f}{\partial z}.\end{aligned}$$

Exercise 4.9 The results are all zero by expansion in Cartesian frame.

Exercise 4.10 Use expansion in Cartesian frame.

Exercise 4.11 Use expansion in Cartesian frame.

Exercise 4.12 Equation (4.18) gives, for instance,

$$\Gamma_{221} = \frac{1}{2} \left(\frac{\partial g_{12}}{\partial q^2} + \frac{\partial g_{21}}{\partial q^2} - \frac{\partial g_{22}}{\partial q^1} \right).$$

In the cylindrical system this is

$$\Gamma_{221} = \frac{1}{2} \left(\frac{\partial 0}{\partial \phi} + \frac{\partial 0}{\partial \phi} - \frac{\partial \rho^2}{\partial \rho} \right) = -\rho.$$

Exercise 4.13 Use the results of the previous exercise and equation (4.20).

Exercise 4.14 From $\mathbf{r} = \hat{\mathbf{x}} c \cosh u \cos v + \hat{\mathbf{y}} c \sinh u \sin v$ we find

$$\begin{aligned}\mathbf{r}_u &= \frac{\partial \mathbf{r}}{\partial u} = \hat{\mathbf{x}} c \sinh u \cos v + \hat{\mathbf{y}} c \cosh u \sin v, \\ \mathbf{r}_v &= \frac{\partial \mathbf{r}}{\partial v} = -\hat{\mathbf{x}} c \cosh u \sin v + \hat{\mathbf{y}} c \sinh u \cos v,\end{aligned}$$

and consequently

$$g_{uu} = g_{vv} = c^2(\cosh^2 u - \cos^2 v), \qquad g_{uv} = g_{vu} = 0$$

(the system is orthogonal). Then

$$\mathbf{r}^u = \frac{\mathbf{r}_u}{c^2(\cosh^2 u - \cos^2 v)}, \qquad \mathbf{r}^v = \frac{\mathbf{r}_v}{c^2(\cosh^2 u - \cos^2 v)},$$

from which we get

$$g^{uu} = g^{vv} = \frac{1}{c^2(\cosh^2 u - \cos^2 v)}, \qquad g^{uv} = g^{vu} = 0.$$

An application of (4.18) gives

$$\begin{aligned}
\Gamma_{uuu} &= c^2 \cosh u \sinh u, \\
\Gamma_{uuv} &= -c^2 \cos v \sin v, \\
\Gamma_{uvu} &= \Gamma_{vuu} = c^2 \cos v \sin v, \\
\Gamma_{vvu} &= -c^2 \cosh u \sinh u, \\
\Gamma_{vuv} &= \Gamma_{uvv} = c^2 \cosh u \sinh u, \\
\Gamma_{vvv} &= c^2 \cos v \sin v.
\end{aligned}$$

Finally, from (4.20) we obtain

$$\begin{aligned}
\Gamma^u_{uu} &= \frac{\cosh u \sinh u}{\cosh^2 u - \cos^2 v}, \\
\Gamma^v_{uu} &= -\frac{\cos v \sin v}{\cosh^2 u - \cos^2 v}, \\
\Gamma^u_{uv} &= \Gamma^u_{vu} = \frac{\cos v \sin v}{\cosh^2 u - \cos^2 v}, \\
\Gamma^u_{vv} &= -\frac{\cosh u \sinh u}{\cosh^2 u - \cos^2 v}, \\
\Gamma^v_{vu} &= \Gamma^v_{uv} = \frac{\cosh u \sinh u}{\cosh^2 u - \cos^2 v}, \\
\Gamma^v_{vv} &= \frac{\cos v \sin v}{\cosh^2 u - \cos^2 v}.
\end{aligned}$$

Exercise 4.15 We switch to a notation in which the coordinate symbols ρ and ϕ are used for the indices. The only nonzero Christoffel symbols are $\Gamma^\rho_{\phi\phi} = -\rho$, $\Gamma^\phi_{\rho\phi} = 1/\rho$. Then

$$\nabla_\rho f^\rho = \frac{\partial f^\rho}{\partial \rho}, \qquad \nabla_\phi f^\rho = \frac{\partial f^\rho}{\partial \phi} - \rho f^\phi,$$

$$\nabla_\rho f^\phi = \frac{\partial f^\phi}{\partial \rho} + \frac{1}{\rho} f^\phi, \qquad \nabla_\phi f^\phi = \frac{\partial f^\phi}{\partial \phi} + \frac{1}{\rho} f^\rho.$$

Exercise 4.16 Proceed using the same method as shown in the text. For example,

$$\begin{aligned}\frac{\partial}{\partial q^k}(a_i^{\cdot j}\mathbf{r}^i\mathbf{r}_j) &= \frac{\partial a_i^{\cdot j}}{\partial q^k}\mathbf{r}^i\mathbf{r}_j + a_i^{\cdot j}\left(\frac{\partial \mathbf{r}^i}{\partial q^k}\mathbf{r}_j + \mathbf{r}^i\frac{\partial \mathbf{r}_j}{\partial q^k}\right) \\ &= \frac{\partial a_i^{\cdot j}}{\partial q^k}\mathbf{r}^i\mathbf{r}_j + a_i^{\cdot j}\left(-\Gamma_{kt}^i\mathbf{r}^t\mathbf{r}_j + \mathbf{r}^i\Gamma_{jk}^t\mathbf{r}_t\right) \\ &= \frac{\partial a_i^{\cdot j}}{\partial q^k}\mathbf{r}^i\mathbf{r}_j + \left(-a_s^{\cdot j}\Gamma_{ki}^s\mathbf{r}^i\mathbf{r}_j + \mathbf{r}^i a_i^{\cdot s}\Gamma_{sk}^j\mathbf{r}_j\right) \\ &= \left(\frac{\partial a_i^{\cdot j}}{\partial q^k} - \Gamma_{ki}^s a_s^{\cdot j} + \Gamma_{sk}^j a_i^{\cdot s}\right)\mathbf{r}^i\mathbf{r}_j.\end{aligned}$$

Exercise 4.17 Consider, for example,

$$\nabla_k g_{ij} = \frac{\partial g_{ij}}{\partial q^k} - \Gamma_{ki}^s g_{sj} - \Gamma_{kj}^s g_{is}.$$

By (4.19) this is

$$\nabla_k g_{ij} = \frac{\partial g_{ij}}{\partial q^k} - \Gamma_{kij} - \Gamma_{kji}.$$

Elimination of Γ_{kij} and Γ_{kji} via (4.18) shows that $\nabla_k g_{ij} = 0$.

Exercise 4.18 Equate components of the first and third members of the trivial identity $\nabla(\mathbf{E}\cdot\mathbf{a}) = \nabla\mathbf{a} = \mathbf{E}\cdot\nabla\mathbf{a}$.

Exercise 4.19 Cylindrical:

$$\begin{aligned}\nabla\cdot\mathbf{f} &= \frac{1}{\sqrt{g}}\frac{\partial}{\partial q^i}(\sqrt{g}f^i) \\ &= \frac{1}{\rho}\left[\frac{\partial}{\partial\rho}(\rho f^1) + \frac{\partial}{\partial\phi}(\rho f^2) + \frac{\partial}{\partial z}(\rho f^3)\right] \\ &= \frac{1}{\rho}\frac{\partial}{\partial\rho}(\rho f^1) + \frac{1}{\rho}\frac{\partial}{\partial\phi}(\rho f^2) + \frac{\partial}{\partial z}(f^3) \\ &= \frac{1}{\rho}\frac{\partial}{\partial\rho}(\rho f_\rho) + \frac{1}{\rho}\frac{\partial f_\phi}{\partial\phi} + \frac{\partial f_z}{\partial z}.\end{aligned}$$

Spherical:

$$\begin{aligned}\nabla\cdot\mathbf{f} &= \frac{1}{\sqrt{g}}\frac{\partial}{\partial q^i}(\sqrt{g}f^i)\\ &= \frac{1}{r^2\sin\theta}\left[\frac{\partial}{\partial r}(r^2\sin\theta f^1)+\frac{\partial}{\partial\theta}(r^2\sin\theta f^2)+\frac{\partial}{\partial\phi}(r^2\sin\theta f^3)\right]\\ &= \frac{1}{r^2\sin\theta}\left[\frac{\partial}{\partial r}(r^2\sin\theta f_r)+\frac{\partial}{\partial\theta}(r\sin\theta f_\theta)+\frac{\partial}{\partial\phi}(rf_\phi)\right]\\ &= \frac{1}{r^2}\frac{\partial}{\partial r}(r^2 f_r)+\frac{1}{r\sin\theta}\frac{\partial}{\partial\theta}(\sin\theta f_\theta)+\frac{1}{r\sin\theta}\frac{\partial f_\phi}{\partial\phi}.\end{aligned}$$

Exercise 4.20

$$\nabla\times\nabla f = \epsilon^{ijk}\mathbf{r}_k\left(\frac{\partial^2 f}{\partial q^i\partial q^j}-\Gamma^n_{ij}\frac{\partial f}{\partial q^n}\right)=\mathbf{0}.$$

(Swapping two adjacent indices on ϵ^{ijk} causes a sign change; hence, whenever the symbol multiplies an expression that is symmetric in two of its subscripts, the result is zero.)

Exercise 4.22 We have

$$\begin{aligned}\left|\frac{\partial(x^1,x^2,x^3)}{\partial(\tilde q^1,\tilde q^2,\tilde q^3)}\right| &= \begin{vmatrix}\frac{\partial x^1}{\partial\tilde q^1} & \frac{\partial x^1}{\partial\tilde q^2} & \frac{\partial x^1}{\partial\tilde q^3}\\ \frac{\partial x^2}{\partial\tilde q^1} & \frac{\partial x^2}{\partial\tilde q^2} & \frac{\partial x^2}{\partial\tilde q^3}\\ \frac{\partial x^3}{\partial\tilde q^1} & \frac{\partial x^3}{\partial\tilde q^2} & \frac{\partial x^3}{\partial\tilde q^3}\end{vmatrix} = \begin{vmatrix}\frac{\partial x^1}{\partial q^i}\frac{\partial q^i}{\partial\tilde q^1} & \frac{\partial x^1}{\partial q^i}\frac{\partial q^i}{\partial\tilde q^2} & \frac{\partial x^1}{\partial q^i}\frac{\partial q^i}{\partial\tilde q^3}\\ \frac{\partial x^2}{\partial q^i}\frac{\partial q^i}{\partial\tilde q^1} & \frac{\partial x^2}{\partial q^i}\frac{\partial q^i}{\partial\tilde q^2} & \frac{\partial x^2}{\partial q^i}\frac{\partial q^i}{\partial\tilde q^3}\\ \frac{\partial x^3}{\partial q^i}\frac{\partial q^i}{\partial\tilde q^1} & \frac{\partial x^3}{\partial q^i}\frac{\partial q^i}{\partial\tilde q^2} & \frac{\partial x^3}{\partial q^i}\frac{\partial q^i}{\partial\tilde q^3}\end{vmatrix}\\ &= \left|\begin{pmatrix}\frac{\partial x^1}{\partial q^1} & \frac{\partial x^1}{\partial q^2} & \frac{\partial x^1}{\partial q^3}\\ \frac{\partial x^2}{\partial q^1} & \frac{\partial x^2}{\partial q^2} & \frac{\partial x^2}{\partial q^3}\\ \frac{\partial x^3}{\partial q^1} & \frac{\partial x^3}{\partial q^2} & \frac{\partial x^3}{\partial q^3}\end{pmatrix}\begin{pmatrix}\frac{\partial q^1}{\partial\tilde q^1} & \frac{\partial q^1}{\partial\tilde q^2} & \frac{\partial q^1}{\partial\tilde q^3}\\ \frac{\partial q^2}{\partial\tilde q^1} & \frac{\partial q^2}{\partial\tilde q^2} & \frac{\partial q^2}{\partial\tilde q^3}\\ \frac{\partial q^3}{\partial\tilde q^1} & \frac{\partial q^3}{\partial\tilde q^2} & \frac{\partial q^3}{\partial\tilde q^3}\end{pmatrix}\right|.\end{aligned}$$

The result follows from the fact that the determinant of a product equals the product of the determinants.

Exercise 4.23 Show that

(1) $\|\mathbf{A}^n\|\le q^n\to 0$ as $n\to\infty$;
(2) $(\mathbf{E}-\mathbf{A})(\mathbf{E}+\mathbf{A}+\mathbf{A}^2+\mathbf{A}^3+\cdots+\mathbf{A}^n)=\mathbf{E}-\mathbf{A}^{n+1}$;
(3) limit passage may be justified in the last equality.

Chapter 5

Exercise 5.1

(a) $\mathbf{r}'(t) = -\mathbf{i}_1 \sin t + \mathbf{i}_2 \cos t + \mathbf{i}_3$, $|\mathbf{r}'(t)| = \sqrt{2}$, so the required length is

$$\int_0^{2\pi} \sqrt{2}\, dt = 2\pi\sqrt{2}.$$

(b) The ellipse can be described as the locus of the tip of the vector

$$\mathbf{r}(t) = \mathbf{i}_1 A \cos t + \mathbf{i}_2 B \sin t \qquad (0 \le t < 2\pi).$$

Hence

$$s = \int_0^{2\pi} (A^2 \sin^2 t + B^2 \cos^2 t)^{1/2}\, dt.$$

The integral on the right is an elliptic integral and cannot be evaluated in closed form.

Exercise 5.3 $s(t) = \int_0^t \sqrt{2}\, dt = t\sqrt{2}$, so

$$\mathbf{r}(s) = \mathbf{i}_1 \cos(s/\sqrt{2}) + \mathbf{i}_2 \sin(s/\sqrt{2}) + \mathbf{i}_3 (s/\sqrt{2}).$$

The unit tangent is

$$\mathbf{r}'(s) = \frac{1}{\sqrt{2}} \left[-\mathbf{i}_1 \sin\left(\frac{s}{\sqrt{2}}\right) + \mathbf{i}_2 \cos\left(\frac{s}{\sqrt{2}}\right) + \mathbf{i}_3 \right].$$

Exercise 5.4 The curve is an exponential spiral. We have

$$\mathbf{r}'(t) = e^t[\mathbf{i}_1(\cos t - \sin t) + \mathbf{i}_2(\cos t + \sin t)],$$

hence $\mathbf{r}'(\pi/4) = \mathbf{i}_2 e^{\pi/4}\sqrt{2}$. Also

$$\mathbf{r}(\pi/4) = \frac{\sqrt{2}}{2} e^{\pi/4}(\mathbf{i}_1 + \mathbf{i}_2),$$

so the tangent line is described by

$$x = \frac{\sqrt{2}}{2} e^{\pi/4}.$$

Exercise 5.5 Assuming the curve can be expressed in the form $\rho = \rho(\theta)$, we can write the position vector as

$$\mathbf{r}(\theta) = \mathbf{i}_1 \rho(\theta)\cos\theta + \mathbf{i}_2 \rho(\theta)\sin\theta.$$

Differentiation gives

$$\mathbf{r}'(\theta) = \mathbf{i}_1[-\rho(\theta)\sin\theta + \rho'(\theta)\cos\theta] + \mathbf{i}_2[\rho(\theta)\cos\theta + \rho'(\theta)\sin\theta].$$

(a) The equation $\mathbf{r} = \mathbf{r}(\theta_0) + \lambda\mathbf{r}'(\theta_0)$ gives upon substitution and matching of coefficients

$$\rho\cos\theta = \rho(\theta_0)\cos\theta_0 + \lambda[-\rho(\theta_0)\sin\theta_0 + \rho'(\theta_0)\cos\theta_0],$$
$$\rho\sin\theta = \rho(\theta_0)\sin\theta_0 + \lambda[\rho(\theta_0)\cos\theta_0 + \rho'(\theta_0)\sin\theta_0],$$

where (ρ,θ) locates a point on the tangent line. The desired analogues of (5.1) can be obtained by (1) squaring and adding, whereby we eliminate θ from the left-hand side, and (2) dividing the two equations, whereby we eliminate ρ from the left-hand side. Elimination of λ instead gives

$$\frac{\rho\cos\theta - \rho(\theta_0)\cos\theta_0}{-\rho(\theta_0)\sin\theta_0 + \rho'(\theta_0)\cos\theta_0} = \frac{\rho\sin\theta - \rho(\theta_0)\sin\theta_0}{\rho(\theta_0)\cos\theta_0 + \rho'(\theta_0)\sin\theta_0}$$

for the analogue of (5.2).

(b) We compute $\mathbf{r}(\theta)\cdot\mathbf{r}'(\theta)$ to get

$$\mathbf{r}(\theta)\cdot\mathbf{r}'(\theta) = \rho(\theta)\rho'(\theta).$$

We also find

$$|\mathbf{r}(\theta)| = \rho(\theta), \qquad |\mathbf{r}'(\theta)| = [\rho^2(\theta) + \rho'^2(\theta)]^{1/2},$$

so that by definition of the dot product

$$\cos\phi = \frac{\mathbf{r}(\theta)\cdot\mathbf{r}'(\theta)}{|\mathbf{r}(\theta)|\,|\mathbf{r}'(\theta)|} = \frac{\rho'(\theta)}{[\rho^2(\theta) + \rho'^2(\theta)]^{1/2}}.$$

Use of the identity

$$\tan\phi = \frac{[1-\cos^2\phi]^{1/2}}{\cos\phi}$$

allows us to get

$$\tan\phi = \frac{\rho(\theta)}{\rho'(\theta)}.$$

Exercise 5.7 Let $\mathbf{r}'(t_0) = 0$. For determining a tangent vector at t_0 we can use the Taylor expansion of $\mathbf{r} = \mathbf{r}(t)$:

$$\mathbf{r}(t_0 + \Delta t) = \mathbf{r}(t_0) + \mathbf{r}'(t_0)\Delta t + \\ + \frac{1}{2!}\mathbf{r}''(t_0)(\Delta t)^2 + \cdots + \frac{1}{n!}\mathbf{r}^{(n)}(t_0)(\Delta t)^n + o(|\Delta t|^n).$$

If the nth derivative is the first that is not zero at $t = t_0$ then

$$\mathbf{r}(t_0 + \Delta t) = \mathbf{r}(t_0) + \frac{1}{n!}\mathbf{r}^{(n)}(t_0)(\Delta t)^n + o(|\Delta t|^n).$$

When $\Delta t \to 0$ the direction of vector $\mathbf{r}(t_0 + \Delta t) - \mathbf{r}(t_0)$ tends to the tangential direction of the curve at $t = t_0$. Thus a tangent vector $\mathbf{t}$ to the curve at $t = t_0$ can be found as

$$\mathbf{t} = \lim_{\Delta t \to 0} \frac{\mathbf{r}(t_0 + \Delta t) - \mathbf{r}(t_0)}{(\Delta t)^n} = \frac{1}{n!}\mathbf{r}^{(n)}(t_0).$$

Exercise 5.8

$$x = x(t_0) + \lambda \frac{1}{n!} x^{(n)}(t_0),$$
$$y = y(t_0) + \lambda \frac{1}{n!} y^{(n)}(t_0),$$
$$z = z(t_0) + \lambda \frac{1}{n!} z^{(n)}(t_0).$$

Exercise 5.9 This time $s(t) = t\sqrt{\alpha^2 + \beta^2}$, hence

$$\mathbf{r}(s) = \mathbf{i}_1 \alpha \cos\left(\frac{s}{\sqrt{\alpha^2 + \beta^2}}\right) + \mathbf{i}_2 \alpha \sin\left(\frac{s}{\sqrt{\alpha^2 + \beta^2}}\right) + \mathbf{i}_3 \beta \frac{s}{\sqrt{\alpha^2 + \beta^2}}.$$

This gives

$$\mathbf{r}'(s) = \frac{1}{\sqrt{\alpha^2 + \beta^2}} \Bigg[- \mathbf{i}_1 \alpha \sin\left(\frac{s}{\sqrt{\alpha^2 + \beta^2}}\right) + \\ + \mathbf{i}_2 \alpha \cos\left(\frac{s}{\sqrt{\alpha^2 + \beta^2}}\right) + \mathbf{i}_3 \beta \Bigg],$$

$$\mathbf{r}''(s) = \frac{\alpha}{\alpha^2 + \beta^2} \left[-\mathbf{i}_1 \cos\left(\frac{s}{\sqrt{\alpha^2 + \beta^2}}\right) - \mathbf{i}_2 \sin\left(\frac{s}{\sqrt{\alpha^2 + \beta^2}}\right) \right],$$

and hence

$$k = |\mathbf{r}''(s)| = \frac{\alpha}{\alpha^2 + \beta^2}.$$

Exercise 5.10 The principal normal is

$$\boldsymbol{\nu} = -\mathbf{i}_1 \cos\left(\frac{s}{\sqrt{\alpha^2+\beta^2}}\right) - \mathbf{i}_2 \sin\left(\frac{s}{\sqrt{\alpha^2+\beta^2}}\right).$$

The binormal is obtained as

$$\boldsymbol{\beta} = \begin{vmatrix} \mathbf{i}_1 & \mathbf{i}_2 & \mathbf{i}_3 \\ -\frac{\alpha}{\sqrt{\alpha^2+\beta^2}} \sin\left(\frac{s}{\sqrt{\alpha^2+\beta^2}}\right) & \frac{\alpha}{\sqrt{\alpha^2+\beta^2}} \cos\left(\frac{s}{\sqrt{\alpha^2+\beta^2}}\right) & \frac{\beta}{\sqrt{\alpha^2+\beta^2}} \\ -\cos\left(\frac{s}{\sqrt{\alpha^2+\beta^2}}\right) & -\sin\left(\frac{s}{\sqrt{\alpha^2+\beta^2}}\right) & 0 \end{vmatrix}$$

and is

$$\boldsymbol{\beta} = \frac{1}{\sqrt{\alpha^2+\beta^2}} \left[\mathbf{i}_1 \beta \sin\left(\frac{s}{\sqrt{\alpha^2+\beta^2}}\right) - \mathbf{i}_2 \beta \cos\left(\frac{s}{\sqrt{\alpha^2+\beta^2}}\right) + \mathbf{i}_3 \alpha\right].$$

We can determine a plane in space by specifying the normal to the plane and a single point through which the plane passes. If the plane passes through the point located by the position vector $\mathbf{r}_0$, and $\mathbf{N}$ is any normal vector, then the plane is described by the vector equation

$$\mathbf{N} \cdot (\mathbf{r} - \mathbf{r}_0) = 0$$

(that is, all points whose position vectors $\mathbf{r}$ satisfy the above equation will lie in the plane). Hence the equations for the three fundamental planes are

$$\begin{aligned} \boldsymbol{\beta} \cdot (\mathbf{r} - \mathbf{r}_0) &= 0 \qquad && \text{osculating plane,} \\ \boldsymbol{\tau} \cdot (\mathbf{r} - \mathbf{r}_0) &= 0 && \text{normal plane,} \\ \boldsymbol{\nu} \cdot (\mathbf{r} - \mathbf{r}_0) &= 0 && \text{rectifying plane.} \end{aligned}$$

The rectifying plane for the helix is found to be $x \cos t_0 + y \sin t_0 = \alpha$.

Exercise 5.11 Writing $\mathbf{r} = \mathbf{r}(t)$, we differentiate with respect to t and use the chain rule to obtain

$$\frac{d\mathbf{r}}{dt} = \boldsymbol{\tau} \frac{ds}{dt}, \qquad \frac{d^2\mathbf{r}}{dt^2} = \boldsymbol{\tau} \frac{d^2 s}{dt^2} + \boldsymbol{\nu} k \left(\frac{ds}{dt}\right)^2.$$

These give

$$\frac{d\mathbf{r}}{dt} \times \frac{d^2\mathbf{r}}{dt^2} = \beta k \left(\frac{ds}{dt}\right)^3$$

or

$$\mathbf{r}'(t) \times \mathbf{r}''(t) = \beta k |\mathbf{r}'(t)|^3.$$

Dotting this equation to itself we obtain

$$k^2 = \frac{[\mathbf{r}'(t) \times \mathbf{r}''(t)]^2}{|\mathbf{r}'(t)|^6}.$$

The formula in the text was presented in such a way that the absolute value function (with its inherent discontinuity at the origin) would be avoided.

Exercise 5.12

(a) $R = 5\sqrt{10}/3$. $R = \infty$.
(b) $x = -b/2a$, the vertex of the parabola.

Exercise 5.15 No; construct counterexamples.

Exercise 5.16 First show that $\boldsymbol{\beta}$ is a constant $\boldsymbol{\beta}_0$. Then use the definition of a plane as a set of points described by a radius vector $\mathbf{r}$ such that $\mathbf{r}\cdot\mathbf{n} = c$, where c is a constant.

Exercise 5.17 Straightforward using equations (5.6) and (5.9).

Exercise 5.18 The vectors $\mathbf{r}'(t)$ and $\mathbf{r}'''(t)$ reverse their directions, while $\mathbf{r}''(t)$ does not. So $\boldsymbol{\tau}$ reverses direction, $\boldsymbol{\nu}$ maintains its direction, and therefore $\boldsymbol{\beta}$ reverses direction. The moving trihedron changes its handedness. The formulas for k_1 and k_2 show that these quantities do not change sign.

Exercise 5.19 We have

$$k_2 = -\frac{\beta}{\alpha^2 + \beta^2}.$$

Note that the ratio of the curvature to the torsion is constant for the helix.

Exercise 5.20

$$\frac{d^3\mathbf{r}}{ds^3} = -k_1^2\boldsymbol{\tau} + \frac{dk_1}{ds}\boldsymbol{\nu} - k_1 k_2 \boldsymbol{\beta}.$$

Exercise 5.21 Straightforward verification.

Exercise 5.22 We have

$$\mathbf{r}'(t) = \mathbf{i}_1 \sin t + \mathbf{i}_2 + \mathbf{i}_3 \cos t, \qquad |\mathbf{r}'(t)| = \sqrt{2},$$

and can find the length parameter for the curve by setting

$$s(t) = \int_0^t |\mathbf{r}'(x)|\, dx = \int_0^t \sqrt{2}\, dx = \sqrt{2}t.$$

So $t = s/\sqrt{2}$ and we have

$$\mathbf{r}(s) = \mathbf{i}_1 \left[1 - \cos\left(\frac{s}{\sqrt{2}}\right)\right] + \mathbf{i}_2 \left(\frac{s}{\sqrt{2}}\right) + \mathbf{i}_3 \sin\left(\frac{s}{\sqrt{2}}\right),$$

$$\mathbf{r}'(s) = \mathbf{i}_1 \frac{1}{\sqrt{2}} \sin\left(\frac{s}{\sqrt{2}}\right) + \mathbf{i}_2 \left(\frac{1}{\sqrt{2}}\right) + \mathbf{i}_3 \frac{1}{\sqrt{2}} \cos\left(\frac{s}{\sqrt{2}}\right),$$

$$\mathbf{r}''(s) = \mathbf{i}_1 \frac{1}{2} \cos\left(\frac{s}{\sqrt{2}}\right) - \mathbf{i}_3 \frac{1}{2} \sin\left(\frac{s}{\sqrt{2}}\right).$$

The desired decomposition of the acceleration is

$$\mathbf{a} = s''(t)\boldsymbol{\tau} + k_1 v^2 \boldsymbol{\nu}.$$

In the present case we have $s''(t) = 0$, while $k_1 = |\mathbf{r}''(s)| = 1/2$ and $v^2 = (ds/dt)^2 = 2$. Therefore $\mathbf{a} = \boldsymbol{\nu}$.

Exercise 5.23 The sphere is described by

$$\mathbf{r} = \hat{\mathbf{x}} a \sin\theta \cos\phi + \hat{\mathbf{y}} a \sin\theta \sin\phi + \hat{\mathbf{z}} a \cos\theta.$$

We have

$$\frac{\partial \mathbf{r}}{\partial \theta} = \hat{\mathbf{x}} a \cos\theta \cos\phi + \hat{\mathbf{y}} a \cos\theta \sin\phi - \hat{\mathbf{z}} a \sin\theta,$$

$$\frac{\partial \mathbf{r}}{\partial \phi} = -\hat{\mathbf{x}} a \sin\theta \sin\phi + \hat{\mathbf{y}} a \sin\theta \cos\phi,$$

hence

$$E = a^2, \qquad F = 0, \qquad G = a^2 \sin^2\theta,$$

so that $(ds)^2 = a^2 (d\theta)^2 + a^2 \sin^2\theta (d\phi)^2$.

Exercise 5.24

(a) The differentials of coordinates for elementary curves on the coordinate lines are $(du^1, du^2 = 0)$ and $(du^1 = 0, du^2)$. The answer is given by $\cos\theta = F/\sqrt{EG}$.

(b) The desired angle is $\psi = \pi/4$.

Exercise 5.25 We find

$$\mathbf{r}_1 = \mathbf{i}_1 \cos u^2 + \mathbf{i}_2 \sin u^2 + \mathbf{i}_3,$$
$$\mathbf{r}_2 = -\mathbf{i}_1 u^1 \sin u^2 + \mathbf{i}_2 u^1 \cos u^2,$$

hence $E = 2$, $F = 0$, $G = (u^1)^2$. Then

$$S = \int_0^{2\pi} \int_0^a \sqrt{2} u^1 \, du^1 \, du^2 = \pi\sqrt{2} a^2.$$

Exercise 5.26 Computing $d\mathbf{r}/d\rho$ and $d\mathbf{r}/d\phi$, we find that

$$E = 1 + [f'(\rho)]^2, \qquad F = 0, \qquad G = \rho^2.$$

From these we get $(ds)^2 = \{1 + [f'(\rho)]^2\}(d\rho)^2 + \rho^2(d\phi)^2$ and

$$S = \int_A \sqrt{EG - F^2} \, d\rho \, d\phi = \iint \sqrt{1 + [f'(\rho)]^2} \rho \, d\rho \, d\phi.$$

Exercise 5.29 Straightforward.

Exercise 5.30 From $\mathbf{r} = \hat{\mathbf{x}}\rho\cos\phi + \hat{\mathbf{y}}\rho\sin\phi + \hat{\mathbf{z}} f(\rho)$ we find that

$$\mathbf{n} = \frac{\frac{\partial \mathbf{r}}{\partial \rho} \times \frac{\partial \mathbf{r}}{\partial \phi}}{\left| \frac{\partial \mathbf{r}}{\partial \rho} \times \frac{\partial \mathbf{r}}{\partial \phi} \right|} = \frac{-\hat{\mathbf{x}} f'(\rho)\cos\phi - \hat{\mathbf{y}} f'(\rho)\sin\phi + \hat{\mathbf{z}}}{\sqrt{1 + [f'(\rho)]^2}}$$

and

$$\frac{\partial^2 \mathbf{r}}{\partial \rho^2} = \hat{\mathbf{z}} f''(\rho), \quad \frac{\partial^2 \mathbf{r}}{\partial \rho \partial \phi} = -\hat{\mathbf{x}}\sin\phi + \hat{\mathbf{y}}\cos\phi, \quad \frac{\partial^2 \mathbf{r}}{\partial \phi^2} = -\hat{\mathbf{x}}\rho\cos\phi - \hat{\mathbf{y}}\rho\sin\phi.$$

Hence

$$L = \frac{f''(\rho)}{\sqrt{1 + [f'(\rho)]^2}}, \qquad M = 0, \qquad N = \frac{\rho f'(\rho)}{\sqrt{1 + [f'(\rho)]^2}}.$$

Exercise 5.31 In an earlier exercise we found that $E = a^2$, $F = 0$, and $G = a^2 \sin^2\theta$ for the sphere. The outward normal to the sphere is

$$\mathbf{n} = \hat{\mathbf{x}} \sin\theta\cos\phi + \hat{\mathbf{y}} \sin\theta\sin\phi + \hat{\mathbf{z}}\cos\theta,$$

and this yields the values $L = -a$, $M = 0$, $N = -a\sin^2\theta$. The mean curvature is $H = -1/a$, and the Gaussian curvature is $K = 1/a^2$.

Exercise 5.33 Straightforward.

Exercise 5.35 The answers include

$$\mu_1^{\cdot 1} = \frac{-f''(\rho)}{\{1 + [f'(\rho)]^2\}^{3/2}}, \qquad \mu_1^{\cdot 2} = \mu_2^{\cdot 1} = 0, \qquad \mu_2^{\cdot 2} = \frac{-f'(\rho)}{\rho\sqrt{1 + [f'(\rho)]^2}},$$

and

$$\Gamma_{11}^1 = \frac{f'(\rho)f''(\rho)}{1 + [f'(\rho)]^2}.$$

Exercise 5.40 A logarithmic spiral

$$\mathbf{r} = ce^{a\phi}[\mathbf{i}\cos(\phi + \phi_0) + \mathbf{j}\sin(\phi + \phi_0)]$$

where the constant c is defined by the initial values of the curve.

Bibliography

Aris, R. *Vectors, Tensors and the Basic Equations of Fluid Mechanics.* Dover Publications, New York, 1990.

Borg, S. *Matrix-Tensor Methods in Continuum Mechanics.* World Scientific, Teaneck, NJ, 1990.

Danielson, D.A. *Vectors and Tensors in Engineering and Physics.* Addison–Wesley, New York, 1992.

Goodbody, A.M. *Cartesian Tensors, with Applications to Mechanics, Fluid Mechanics, and Elasticity.* Ellis Horwood, 1982.

Jaeger, L. *Cartesian Tensors in Engineering Science.* Pergamon Press, Oxford, UK, 1966.

Jeffreys, H. *Cartesian Tensors.* Cambridge University Press, Cambridge, UK, 1931.

Joshi, A. *Matrices and Tensors in Physics.* Wiley, New York, 1995.

Knowles, J. *Linear Vector Spaces and Cartesian Tensors.* Oxford, 1997.

Lurie, A.I. *Non-linear Theory of Elasticity.* North-Holland Series in Applied Mathematics and Mechanics, 36, 1990.

McConnell, A. *Application of Tensor Analysis.* Dover Publications, New York, 1957.

O'Neill, B. *Elementary Differential Geometry.* Academic Press, New York, 1997.

Papastavridis, J. *Tensor Calculus and Analytical Dynamics.* CRC Press, Boca Raton, FL, 1998.

Pogorelov, A.V. *Differential Geometry.* Translated from the first Russian ed. by L.F. Boron. P. Noordhoff, Groningen, 1957.

Schouten, J.A. *Tensor Analysis for Physicists.* Clarendon Press, Oxford, 1951.

Sokolnikoff, I.S. *Tensor Analysis: Theory and Applications to Geometry and Mechanics of Continua.* Wiley, New York, 1994.

Synge, J., and Schild, A. *Tensor Calculus.* Dover Publications, New York, 1978.

Talpaert, Y. *Differential Geometry with Applications to Mechanics and Physics.* Marcel Dekker, New York, 2000.

Truesdell, C. *A First Course in Rational Continuum Mechanics.* Academic Press, Boston, 1991.

Young, E. *Vector and Tensor Analysis.* Marcel Dekker, New York, 1993.

Index